Lawson Tait

An Essay on Hospital Mortality

Based upon the Statistics of the Hospitals of Great Britain for Fifteen Years

Lawson Tait

An Essay on Hospital Mortality
Based upon the Statistics of the Hospitals of Great Britain for Fifteen Years

ISBN/EAN: 9783337161866

Printed in Europe, USA, Canada, Australia, Japan

Cover: Foto ©berggeist007 / pixelio.de

More available books at **www.hansebooks.com**

AN ESSAY

ON

HOSPITAL MORTALITY

BASED UPON THE STATISTICS

OF THE

HOSPITALS OF GREAT BRITAIN

FOR FIFTEEN YEARS

BY

LAWSON TAIT, F.R.C.S. ED. & ENG.

FELLOW OF THE STATISTICAL SOCIETY, ETC. ETC.

AUTHOR OF "THE PATHOLOGY AND TREATMENT OF DISEASES OF THE OVARY;"
"DISEASES OF WOMEN ;" ETC. ETC.

LONDON
J. & A. CHURCHILL, NEW BURLINGTON STREET
1877

TO

THE MEMORY OF A GREAT MASTER,

JAMES YOUNG SIMPSON,

THIS EFFORT

IS DEDICATED

BY

A GRATEFUL PUPIL.

PREFACE.

The discussion on Hospitalism raised by Sir James Y. Simpson was unfortunately cut short by his death. His papers were placed in my hands by his son, but I found so little that had not been used by him, that I felt that any further investigation of the subject must be prefaced by research in another direction.

The statistics of amputations, upon which Sir James founded most of his argument, have not been disputed as far as those derived from hospitals are concerned. Those which were contributed by private practitioners were, however, manifestly open to the objection that the records might be imperfect; and, at least, that they were not sufficiently extended to be convincing. I have sifted the original returns very carefully, and am quite certain that, save in one instance, they are all above the suspicion of intentional or even of careless misstatement.

The more I thought this difficult subject over, the more I became satisfied that the first step was to establish the facts of a total hospital mortality for a definite and somewhat extended period. In 1871 I made an attempt to collect the details, given in my tables, from every hospital in Great Britain, for the preceding decade; but it will be seen that the results obtained were such as could hardly be depended upon for accuracy, and in some respects they were so meagre as hardly to be worth the trouble involved in collecting them. I do not, therefore, place great reliance on them for my conclusions.

After waiting for another five years to pass, I renewed my efforts with much greater success ; and I venture to think that if I have done nothing more than insure greater accuracy and a larger amount of detail in the published statistics of many hospitals, my labours will not have been altogether in vain.

As to the accuracy of the figures, I can only say that in every case they are given on the authority of some recognised hospital official. As to the conclusions to be drawn from them, I think I may state that I have exercised as much caution as possible. In some instances I have indicated what I think may fairly be inferred ; but I must ask my readers to regard this, as I have done, merely as a preliminary inquiry. Having ascertained the facts of hospital mortality, we have next to inquire into the causes or explanations of excess, and then the remedies will become apparent. I have gathered a large mass of statistics bearing on special lines of inquiry, such as amputation mortality, and these, I think, will surprise others as they have surprised me.

One value the present statistics must unquestionably possess, especially those for the second period, in that they supply us with certain facts hitherto not exactly ascertained, which may be regarded as hospital constants. The enormous mass of figures employed, including over two millions of patients for the two periods, of which more than three-quarters of a million belong to the six years from 1870-75 inclusive, give us complete assurance, for instance, of the value of the figures representing the period of residence in all hospitals, and a comparison of the varying time spent in hospitals of different kinds and sizes.

Finally, I think I may point out that the figures show incontestably that a most stringent inquiry is demanded as to the possibility of reducing the number of deaths in at least some institutions for the treatment of the sick poor.

BIRMINGHAM, *June*, 1877.

HOSPITAL MORTALITY.

In such an essay as this is intended to be, no useful purpose would be served by an antiquarian discourse on the
history of hospitals. As long as men lived in a state in
which the struggle for existence was maintained as an actual
warfare between individuals, it was, of course, alike impossible and impolitic to take care of the sick and wounded.
They suffered the inevitable fate of the unfit. It was only
when development had been carried so far as to introduce
what has been called the social instinct, that it was discovered that it might be useful to lend a helping hand to
the infirm. Indeed, we may say this social instinct must
have been far advanced when this discovery was made, for
we see still existing human races so far advanced in civilisation as to be skilled navigators who yet yield to their primitive instincts of self-preservation so far as to bury their
aged and ailing whilst yet alive, to save what is really the
" expense " of keeping them.

Confining our inquiries to our own race, we find that
the first attempts to help the sick and wounded were made,
not from a charitable, but from an essentially selfish motive.
Religion, or at least what was at the time called religion,
prompted our forefathers to carry on warfare in the East,
and this warfare was of no patriotic or even chivalrous kind,

but simply to secure, or to contribute towards, the personal salvation of those engaged on one side of the struggle. Out of the battles and bloodshed there grew a division of the labour, so that in 1099 the Order of the Knights of St. John was instituted at Jerusalem, for the care of the wounded Crusaders—a service which they undertook, and no doubt carried out as well as they knew how, solely for the future rewards they expected from it. They established a permanent military hospital at Rhodes, and afterwards at Malta, under the patronage of Charles V. When visited by Howard in 1787, the hospital was still under the direction of the Grand Master and his Knights, and was in a state very characteristic of its origin. It had grown to be so large as to hold 520 patients, probably by reason of legacies, the donors of which contributed them for much the same reason as the Knights undertook their charge, and it had ceased to be military. Its immediate governance was always left to the care of one of the youngest and least experienced of the Order, and for his assistants he had only twenty-two incompetent servants. Howard tells us, with that quaintly simple antithesis which becomes sometimes so pathetic in his unadorned writings, that in the Grand Master's stable forty attendants were kept for twenty-six horses, and that the stables were clean and well supplied with water— in these and other respects contrasting most favourably with the human charge of the priestly Knights. The mortality in such a hospital must have been great, but I doubt if it could ever have exceeded the mortality at Scutari in 1855. At all events, Howard tells us that the " slow hospital fever was prevalent."

Hospitals of various kinds had been founded in this and other countries of Europe, for various purposes and from

various motives, previous to the eighteenth century, but it
is really with the foundation of Guy's Hospital in 1722 that
our modern system of aid for the sick may be said to have
taken its origin. It seemed, however, to take about twenty
or thirty years more to persuade well-to-do people that it
was their duty, or perhaps, to speak more plainly, that it
was to their interest to establish hospitals for the treatment
of disease. We therefore find that our oldest city and
county hospitals date generally from about 1750, and in the
great majority of instances that their foundation was due,
as it is to this day, to the exertions of those most deeply
interested in their existence—practitioners of medicine.

It cannot be surprising, if we look at the houses in which
our forefathers lived in the sixteenth and seventeenth centuries, that their children in the eighteenth should be ignorant
of all true principles of hospital construction or management; or that they should be impressed with any other
idea than that which, in Miss Nightingale's words, seemed
to make it " sufficient for all purposes of curing and healing,
that the sick man and the doctor should merely be brought
together, in any locality, or under any condition whatever."
But it is surprising to find that nearly a hundred years
after Howard's vivid descriptions of hospital misconstruction
and mismanagement, and many years after the burning
words of Florence Nightingale, that in a great hospital of
six hundred beds we have been able to diminish the mortality only one per cent. from what it was in Howard's
time. (Guy's Hospital mortality rate from 1780–90,
10·2 per cent.; 1850–60, 9·1 per cent.) If we go further
back still, to the first five years of the existence of Guy's
Hospital, we find the mortality 13·8 per cent. If we also
bear in mind that then there were many zymotic diseases,

now unknown, all of which were treated in the hospital, and almost only there, and that even of those which still remain to us cases are admitted to the hospital only by accident, and in a proportion which is infinitesimal (about ·38 per cent.), the conclusion is inevitable that hospital hygiene has not advanced as it might and ought to have done. It is greatly to be feared that therapeutical, discoveries, and even surgical improvements, have had nothing to do with what little diminution there has been in hospital mortality, but that it is in greatest part to be credited to general hygienic improvements. The slow, snail-like progress of the mortality rate of Guy's Hospital from 13·8 per cent. in 1730 to 9·1 per cent. in 1860 is very confirmatory of this fear. It is really astonishing how slow progress has been in social and domestic government. It is nearly a hundred years since John Howard, with a prescience which seems to me as great as any discoverer has ever exhibited, advocated the performance of executions in private. He used all the arguments which were heard when the " Private Executions Bill" was passed without opposition, and he recorded them in a book which created a profound excitement when published. In his letters published after his death, he gives a picture of a hospital at Crements-chnock, which, however awful it may seem, has had a parallel in our own time. There were six wards thirty-four and a half feet wide, having four rows of beds, about twenty-two in each row, on a barrack, separated only by a board eight or nine inches high, the walk between the rows only eight feet wide. Scurvy and bloody flux abounded, and from a half to a third of all the patients died. At Witowka there was another such, where the barrack bed had no partitions, and the patients lay so close that there were from sixteen to twenty in a space of thirty feet, and each set of blankets had to cover three or four.

Miss Nightingale tells us that in 1855, at the Scutari Hospital, the men were laid on palliasses on the floor as close as they could lie; there were two rows of beds in the Barrack Hospital corridors, where two persons could hardly pass abreast between foot and foot; and that in seven short months we lost a third of our heroic army from disease alone, and that disease of a purely preventible kind, much of it having the same scorbutic and bloody flux character which Howard lamented eighty years before.

This lesson was such a terrible one, and its experimental results excited such popular wrath, that it is never likely to be repeated.

But we have had it urged upon us by Howard, Miss Nightingale, and many others, and last of all by Simpson, that a loss of life as great, though not so striking, is constantly going on in our civil hospitals, and that it may be checked by exactly the same means which in 1855-6 brought the disease death rate of our Crimean army down from 40 per cent. to less than 3 per cent. That there is some truth in this no one who has seen much hospital work can doubt for a moment, though to what extent it is true must, I fear, long remain a mystery; and that chiefly for the reason that it is almost impossible to obtain data which are not open to objections more or less forcible. How carelessly kept are the records of most of our medical charities none know save those who have had to examine them. The managing authorities are usually content with publishing a report which contains a balance-sheet and a bare statement of a number of patients which have been treated during the year, often without mentioning so important a feature as the number of deaths which have taken place. And this is by no means confined to small hospitals, for some of the largest and most important hospitals in the country publish reports which are absolutely worthless as sources of information. In one point they all join. There is a uniform tone of congratulation on the success of the hospital, and an increase of numbers of the patients is hailed with rejoicing, whilst the committees always regret when the " usefulness of the hospital has been somewhat diminished during the past year." Surely this is done in thoughtlessness. They must be oblivious to the fact that any one accepting gratuitous assistance is being pauperised, and that our system of indiscriminate medical relief has much to answer for in the improvidence of our labouring population. Instead of congratulating themselves on their increased usefulness, the hospital authorities should annually express regret either that human misery should be, in spite of our growing wealth and advancing civilisation, so much on the increase, or that they are the means of doing so much harm.*

Be this, however, as it may, one thing will, I think, be admitted on all hands. If any body of men take upon themselves not only to administer public charity, but to look after the lives and health of our poorer neighbours, they are bound to give an account not only of their expen-

---

* In the *Statistical Journal* for March, 1856, Dr. Guy tells us that nearly one-third of the whole population of the parishes of St. Clement Danes (4720 : 15,662) and St. Mary-le-Strand (817 : 2517) apply at King's College Hospital for medical relief. He found, out of 335 males, 230 to be in work, and 105 out of work ; so that he estimates that nearly 20,000 men in the receipt of wages applied for charitable medical relief in one year (1851) at that one hospital. Out of 67 men, he found 52 earning above 20s. a week, 39 earning 25s. a week, and 30 earning 30s. Evidence of similar and even greater abuse is being accumulated on all sides.

diture, but of their results. There are few hospitals who do this completely and well; but amongst those which do, I think it desirable to mention especially, as worthy of imitation, the reports of Charing Cross Hospital and of the Infirmaries of Glasgow, Paisley, and Greenock, and the Hospital for Sick Children in Birmingham.

It certainly is somewhat remarkable that whilst we take the greatest possible care of our pauper, lunatic, and criminal population, we entirely neglect to place any official supervision over the care of our medical charities. If it is found that in any workhouse, asylum, or prison the death-rate rises unusually high, a commissioner or inspector at once visits and reports. But no such care is exercised over a class of the population infinitely more valuable and far more worth caring for than lunatics, paupers, or felons. In the tables which are given afterwards, evidence will be found which, if not conclusive, at least makes it very likely that there are in this country hospitals where the mortality is raised by causes intrinsic and removable. That hospital mortality has been made positively enormous by mismanagement needs no re-statement. That hospital management is yet perfect is by no means clear. That hospital results are not equal is in evidence; and I have no hesitation in asserting that the onus rests upon the committee of every hospital to show that they are doing the best that can be done under their particular circumstances. To render this clear, there are certain data of a uniform kind which should be in every hospital report. To these I shall afterwards allude at length.

In an inquiry into hospital mortality it must be distinctly borne in mind that there are three steps in the process which, although inter-dependent, must be kept distinct.

The first is to ascertain what hospital mortality really is, and this must be done on some uniform and general plan, and, as far as human power can do it, it should be without prejudice. For any one who has been, as I have, associated with hospital work for the greater part of his life, to be entirely free from prejudice is a most difficult matter. We look on hospitals almost as the means of our existence, and to attack them, or do anything to raise adverse criticism against them, looks like medical heresy. It is not to be wondered at, therefore, that Simpson's papers on hospitalism were received with but little favour; and that two eminent hospital officers should have been induced to write an elaborate but very diffuse blue book, which is little better than an apology for hospitals. Nothing yet written on the subject, on the one side or on the other, is sufficiently precise to satisfy the wants of the statistician, and it is by statistics alone that the first of our three steps can be made. It is futile to say that anything can be proved by figures, though it is at the same time partially true. But if figures are examined with the intent of merely seeing in what direction they point, not with a want to establish any particular view, they will infallibly tell the truth. If this is not so, then the admirable reports of the Registrar-General are useless, and the enormous insurance business based upon them is a gigantic commercial fallacy.

Having established what hospital mortality really is, the next step is to inquire into the causes of its fluctuation, and the third is to discover remedies for the defects laid bare.

The main object of this essay is with the first of these steps. I have made an attempt to tabulate hospital mortality on a uniform plan; and if I have not been successful in obtaining exact results, I am quite certain it is not from any want of having taken a vast amount of trouble about it, and exercised every care and caution which I could think of. I am quite certain, at least, that my efforts are less open to objection than any yet made public, and I am perfectly confident that they are as much as possible free from prejudice.

With the second step I can deal only partially—it is so wide a subject. Upon the third, I could not touch without much more extensive data than are yet in my possession.

I have, however, been able to collect material, which is placed in an Appendix, and which may contribute to the elucidation of these points.

Some years ago circumstances led me to take an especial interest in hospital mortality; and with the purpose of gathering material, I sent a circular to every hospital in Britain, asking for certain details. I confined my inquiry to our own hospitals, because to extend it to the Continent or to America would, I knew, be certain to introduce unknown quantities into my calculations. The details I asked for were simple enough, but I had little anticipation of the difficulties I should encounter. I knew that a similar attempt had been made on a smaller scale by a committee of the Statistical Society, but I had no idea how meagre the results were which they were able to obtain. My circular asked for a statement of the number of beds for each year of the decade from 1861 to 1870, the number of in-patients, and the number of deaths. It was sent to over three hundred hospitals, and received an immediate reply from about half. After repeated applications, I managed to get statistics from two hundred and sixty-three, though in many cases I had to extract it from reports, and in a few I availed myself of the material entered in the journal of the Statistical Society. Fifteen hospitals informed me that they had no available statistics, and from the rest I got no replies whatever. Of those hospitals from whose records I obtained information, one hundred and forty-one returns were for the whole ten years, ten were nine years, eleven for eight years, eleven for seven years, eight for six years, nine for five years, twenty-four for four years, twelve for three years, twenty for two years, and seventeen for one year.

I also obtained a large number of reports—in a few instances, complete sets of them, for the decade. From these I discovered that my returns were of but little use, for several reasons, but chiefly for two. First, it became evident that a very common custom exists of counting a number of patients twice, and in some instances even three and four times over.

Thus patients remaining on the books at the end of a hospital year are very often counted along with the fresh cases admitted as making the total of the in-patients for the year. In this way a number of patients are reckoned twice, and I found that in some hospitals it made a difference of nearly ten per cent. of the whole returns.* Then re-admissions on the ticket system are often counted as two or three additional patients. I found that these two plans sometimes made a difference of one per cent. on the patient death rate for the decade—of course, in favour of the

---

* There is also a perplexing custom very common among hospitals, which may be a source of error in statistics, though the error cannot be great if a number of years are employed. Instead of reckoning a year as from the first of January to the thirty-first of December, they fix on some day in any part of the year, which probably represents the anniversary of the opening of the institution. This should be altered, and the year for every hospital should begin on the first of January. The other plan introduces endless confusion in the accounts.

hospital. In the majority of cases this seems to be only a pious fraud to magnify the importance of the work of the charity, but in some instances it almost amounts to deliberate dishonesty, for by dividing the total expenditure for the year by the number of patients thus improperly enlarged, and by contrasting the result with figures taken from other hospitals where the sum of the patients had not been so magnified, certain institutions have been made to appear in an altogether undeserved light. Then, again, from some returns I found that deaths within twenty-four hours of admission had been removed, as if these were not as much part and parcel of the hospital economy as any of the others.

In asking for a return of the beds used in each hospital, my object was to place along side the death rate a figure which would show how the hospital was used; but I found that the returns would yield such information in only a very few instances. It was perfectly evident that the mere death rate of a hospital in which the patients remained, on an average, fifty days, would yield no basis of comparison with one in which they remained only twenty-five or thirty.

I found the same tendency to exaggeration here, for it was not an unusual thing to find a hospital returning two or three times the number of beds which it could, by any possibility, have in actual use.

I therefore found that my statistics, gathered with much labour, were so full of error that I could use them with but little effect; and although I have embodied them in this essay, it is chiefly to use them for purposes of comparison in certain cases where I have been able to insure accuracy, and because in the cases of two classes, the Irish county infirmaries and the hospitals for children, my first set of returns are, strange to say, more complete than my second.

During the last six years a large number of new hospitals have sprung into existence, chiefly belonging to the classes of special or cottage hospitals. My lists therefore include four hundred and thirty-nine hospitals of all kinds, exclusive only of ophthalmic hospitals and a very few others where, from the nature of the practice carried on in them, deaths rarely occur.

Out of this large number two instances only occurred where my application for statistics met with any want of courtesy. The house surgeon of the infirmary of Rochdale refused to give me the information, and the authorities of the Luton Hospital in Bedfordshire returned my circulars and letters without explanation. Six applications to the Middlesex Hospital received no reply. I obtained their reports, however, by the intervention of a friend, but these documents give no returns of deaths nor any other information of great value, and I had to take the deaths given by the Registrar-General in his reports as occurring in the hospital as the basis of my calculations of its death rate. By this the hospital probably suffers somewhat in its comparison with similar institutions, but for this I am not responsible. I hold that the facts of all public institutions are public property, and should be available in a published and authenticated form.

In pleasing contrast to this, I am in a position to express my thanks to the authorities of a very large majority of the hospitals for the courtesy with which they treated my application, and for the great care they took to insure correctness. From a very large number, each of which is marked by an asterisk in the summarised returns, I obtained copies of the hospital reports which enabled me to insure the accuracy of the returns. In only a few instances did I find it necessary to make corrections. From one hundred and twelve hospitals I received no reply to repeated applications, and in these cases it must be concluded either that there was no information to give, no records being kept, or that it was, in the so-called interests of the hospitals, considered not advisable to give them.

From forty-four hospitals, or about 10 per cent. of the whole, I was unable to extract any reply, either in 1871 or in 1876, whilst sixty-seven of those who gave information in 1871 did not reply in 1876, and of all these I think it necessary here to give special lists :—

Forty-four Hospitals from which no Returns could be obtained for either of the periods.

Stated No. of Beds.		Stated No. of Beds.	
City of Dublin	130	Harrow, Middlesex	7
Inverness	120	Bournemouth	4
Cork County	108	Lewes	4
Sir P. Dun's, Dublin	80	Ross, Hereford	
Jersey Infirmary			
St. Mary's, Manchester	50	*Irish County Infirmaries.*	
Ashton-under-Lyne	44	Londonderry	128
Ryde	42	Limerick	100
Maidstone	40	Galway	80
Newark	36	Tipperary	72
Lanark	33	Kilkenny	70
Douglas (Isle of Man)	32	Monaghan	60
Croydon	18	Sligo	60
Kidderminster	28	Kildare	52
Alloa	15	Westmeath	40
Newtown, Montgomery	15		
Aberystwith	14	*Fever.*	
Edinburgh Med. Missionary	13	Leeds	80
Altrincham	12	Londonderry	72
Saltaire	12	Newry, Armagh	30
Alnwick	11	Arklow, Wicklow	10
Penzance	9		
Hatfield, Essex	8	*Children's.*	
Wrexham	8	Pendlebury, Manchester	84
Bideford, Devon	7	Clinical, Manchester	46

*Sixty-seven Hospitals which gave Returns for 1861–70,
but not for 1870–75.*

Stated No. of Beds.		Stated No. of Beds.	
Bristol Royal Infirmary	242	Fairford	8
St. Vincent's, Dublin	100	Monmouth	8
Plymouth	90	Pembroke	8
Jervis Street Hospital, Dublin	80	Tetbury	8
Guernsey Catcl Hospital	60	East Grinstead	7
West London	60	Cranleigh	6
Truro	52	Downton	6
Waterford	50	Driffield	6
St. Bartholomew's, Chatham	48	Hambrook	6
King's Lynn, Norfolk	48	Harrogate	6
Poplar Hospital	48	King's Sutton	6
Limerick City	40	Worksop	5
Metropolitan Free	40	Crimond	4
Stamford	40	Charmouth	3
Torbay	40		
Tunbridge Wells	40	*Irish County Infirmaries.*	
North Ormsby	30	Maryborough	100
Bolton	26	Roscommon	85
Bootle	26	Downpatrick	80
Stratford-on-Avon	23	Wexford	72
Bangor	20	Cavan	70
Ditchingham, Norfolk	20	Clare	60
Balfour Hospital, Kirkwall	18	Mayo	60
Weybread, Suffolk	18	King's County	50
Loughborough	16	Kerry	40
Weymouth	16	Longford	39
Brecknock	14		
Crewkerne	13	*Children's.*	
Newport	12	Evelina	100
Shepton Mallet	12	Liverpool, Myrtle Street	80
Pembrokeshire Infirmary	12	Edinburgh	72
Bromley	10	Bristol	50
Ilfracombe	10	Gloucester	24
Tewkesbury	10	London North Eastern	24
Wallasey	10	Belgrave, London	19
Dinorwic	8		

My circular of 1876 asked for information for the six
years, from 1870 to 1875, and in the great majority the
returns given were for the whole six years. The details

were tabulated for each year separately, and included the following points: the average number of beds occupied, or the average daily population of the hospitals, the total number of in-patients admitted, the average residence in days, and the number of deaths.

It will be seen that by asking for the average number of beds occupied and the average residence I was able to make the one column correct the other. For if we suppose a hospital with an average daily population of a hundred, and a total number of admissions of a thousand, it will be evident that, as each bed will have ten occupants during the year, the average residence will be 36·5 days. On the converse, if the average residence be given as 36·5 days, and the total number of patients 1000, the average daily population will be 100.

I have given the death rate in two ways—first, in relation to the beds occupied by raising each hospital to the standard of hundreds of beds, and giving the numbers of deaths which each bed or hundreds of beds would have in a year; and, secondly, by a percentage of the patients.

I have arranged the whole number of hospitals in the order of the numbers of beds given in the list of hospitals in Churchill's "Medical Directory," subdividing them into six groups. The first contains the general hospitals; the second contains the Irish county infirmaries, which I have placed by themselves, because I have completely failed to obtain such information concerning them as will throw light upon their peculiar results, and also because they present a special feature in having pecuniary help from the State and subsidies from their counties.

In another group are placed the special zymotic hospitals, and also some special returns of zymotic cases from general hospitals. I have also placed by themselves, in separate groups, lying-in hospitals, hospitals for women, and hospitals for children.

In a special table I have arranged the hospitals in the order of the number of beds in actual occupation, and by an arbitrary division of them into groups I have been able to construct an interesting series of curves.

A great deal of time has been spent in making the necessarily numerous calculations as free from error as possible. As the number of returns for each column is not constant, the divisors have consequently varied.

P.S.—A few returns have been entered since the columns were made up, but none which are at all likely to influence the averages, with the exception of the return for the Birmingham Corporation Small-pox Hospital, which has had such good results as to lower slightly the average hospital death rate of this disease.

GENERAL HOSPITALS, 1861—70.

No.	Name of Hospital.	Years for which return is made.	Full No. of Beds.	Average Beds Occupied.	Average No. of In-patients.	Average No. of Patients to each Bed.	Mean Residence.	Mortality, per cent. of Beds.	Mortality, per cent. of Patients.	District Mortality per 1000.	Ratio of Hosp. to District Mortality.	Remarks.
1	St. Bartholomew's Hosp.	1861-64	676	547·	5489·5	9·86	37·	108·59	10·82	...	...	Statist. Soc. Journal.
2	Guy's	1861-70	600	497·5	4913·6	9·87	36·99	98·15	9·93	...	...	
3	London	1861-70	570	521·8	4478·	8·58	42·54	88·13	10·27	...	...	
4	Roy. Inf. Glasgow	1861-70	547	500·	5730·	11·46	31·93	118·08	10·30	...	...	
5	St. Thomas'	1861-70	211	251·5	2047·5	8·14	44·84	90·37	11·10	...	...	
6	Roy. Inf. Edin.	1861-70	565	395·3	4352·	11·	33·18	125·07	10·96	...	...	
7	St. George's	1861-70	353	345·	3755·	10·88	33·55	101·27	9·30	...	...	
8	Leeds Gen. Inf.	1861-70	200	150·	1682·	11·21	32·56	91·73	8·27	...	...	
9	Middlesex	...	305					...	...	...	...	No information to be obtained.
10	Roy. Inf. Aberdeen	1861-69	300	140·5	2113·	15·	24·33	87·11	6·25	...	...	
11	Manchester Inf.	1861-70	258	216·	2572·4	11·9	30·67	103·48	10·86	...	...	
12	Dundee Inf.	1861-70	260		1940·				8·05	...	...	
13	Liverpool Inf.	1861-70	270	258·	2953·	11·44	31·9	66·89	5·84	...	...	
14	Birmingham Gen.	1861-70	234	168·5	2375·	14·1	26·	119·76	8·49	...	...	
15	Steevens's, Dublin	1861-70	250		2163·				2·77	...	...	
16	Bristol Roy. Inf.	1861-70	242		2583·5				5·59	...	...	
17	Devon and Exeter	1861-70	230		1416·				3·60	...	...	
18	Mater Misericordiæ, Dub.	1867-70	230		565·				6·42	...	...	
19	Newcastle Inf.	...	250							...	...	No reply.
20	Roy. Albert, Devonport	1864-70	218		...				1·86	...	...	About 75 per cent. of beds are Lock.
21	Leicester Inf.	1861-70	200		1506·				5·17	...	...	
22	Wolverhampton Gen.	1861-70	103		718·				8·70	...	...	
23	Liverpool Southern	1861-70	120	817·7	1239·	15·16	24·07	115·54	7·61	...	...	
25	Westminster	1869-70	200		1266·				14·56	...	...	
26	House of Industry, Dublin	1861-70	312		3364·				6·73	...	...	Including Hardwicke Fever.
28	North Stafford Inf.	1861-69	195		1118·				4·98	...	...	

c

GENERAL HOSPITALS, 1861—70 (continued).

No.	Name of Hospital.	Years for which return is made.	Full No. of Beds.	Average Beds Occupied.	Average No. of In-patients.	Average No. of Patients to each Bed.	Mean Residence.	Mortality, per cent. of — Beds.	Mortality, per cent. of — Patients.	District Mortality per 1000.	Ratio of Hosp. to District Mortality.	Remarks.
29	Derby Gen. Inf.	1861-70	150	96·7	977·3	9·9	36·86	58·53	5·79	...	...	Including fever.
30	Paisley Inf.	1861-70	250	...	948·	...	...	...	6·29	...	...	
31	Sussex County	1861-70	165	...	1314·	...	...	...	5·53	...	...	
32	King's College	1861-70	152	...	1607·	...	...	...	11·55	...	...	
33	Radcliffe Inf., Oxford	1861-70	149	...	1108·	...	...	...	3·52	...	...	
34	St. Mary's, Paddington	1861-70	165	152·	1795·	11·8	30·9	115·22	9·76	...	...	
35	Belfast Gen.	1861-70	160	...	1348·	...	...	...	7·43	...	...	
36	Sheffield Gen. Inf.	1861-70	160	...	1165·	...	...	...	9·75	...	...	
37	Bristol Gen.	1861-70	130	...	1306·	...	...	...	5·84	...	...	
38	Charing Cross	1861-70	150	...	1086·	...	...	...	8·81	...	...	
39	Chester Gen. Inf.	1861-70	150	60·	664·	11·06	33·	69·	6·23	...	...	
40	Queen's, Birmingham	1868-70	180	137·	1403·	11·	33·18	49·58	6·79	...	...	
41	University College, Lond.	1861-70	150	116·	1407·	12·13	30·	151·07	14·26	...	...	
42	Hull Gen. Inf.	1861-70	150	101·	1038·5	10·28	35·5	117·93	7·67	...	...	
42a	Bath Gen.	1861-70	145	...	1049·	...	...	...	7·89	...	...	
43	Liverpool Northern	1861-70	146	136·	1405·	10·3	35·43	79·62	7·72	...	...	
44	Nottingham Gen.	1861-70	142	...	1259·	...	...	...	4·75	...	...	
45	Bradford Gen. Inf.	1864-70	120	76·79	744·6	9·56	38·18	62·24	6·42	...	...	
46	Gloucester Gen. Inf.	1861-70	140	...	648·	...	...	...	4·73	...	...	
48	Roy. Berks	1861-70	120	...	860·	...	...	...	4·01	...	...	
49	Salop Inf.	1861-70	140	...	1024·5	...	...	...	5·60	...	...	
50	Northampton Gen. Inf.	1861-70	126	115·5	1370·	11·77	31·	42·18	3·48	...	...	
52	Addenbrooke's, Cambridge	1861-70	120	...	647·5	...	...	...	5·57	...	...	
53	Huddersfield Inf.	1861-70	60	88·	349·	...	...	...	8·83	...	...	
54	Meath, Dublin	1861-70	120	122·	1174·	13·34	29·8	50·	5·07	...	...	
56	Norwich and Norfolk	1861-70	150	...	983·	8·8	41·5	36·	5·65	...	...	
58	Sunderland Gen. Inf.	1861-70	110	...	350·	...	...	...	9·18	...	...	
59	Hants County Inf.	1861-70	108	...	778·	...	...	...	2·96	...	...	
60	Kent and Canterbury	1861-70	120	...	571·5	...	...	...	6·22	...	...	With fever cases.

No.	Hospital	Years								Source
										Statist. Soc. Journal. / Hospital and Workhouse.
61	Dumfries Inf.	1861–70	80	40·	496·	124·	2943	6260	4·65	
63	Adelaide Hosp., Dublin	1861–70	100		8545·				6·36	
64	North Devon Inf., Barnstaple	1861–70	100		615·				2·14	
65	Bedford Gen. Inf.	1861–70	100		709·				2·66	
66	Carlisle Inf.	1861–70	52		479·				4·23	
68	German Hosp., Dalston	1861–70	100	69·	950·	13·76	26·52	10927	7·93	
69	Leamington	1861–70	90		218·				5·27	
70	Lincoln County	1861–70	100		714·5				3·73	
71	Roy. Free, London	1861–4	98	787·	1278·5	14·8	247·	11245	7·	
72	Salisbury Gen. Inf.	1861–70	100		994·				2·97	
74	Worcester Inf.	1861–70	100		1060·				4·33	
75	St. Vincent's, Dublin	1861–70	120		940·				5·59	
76	York County	1861–70	94		6775·				4·09	
77	Stafford Gen. Inf.	1861–70	90		657·				3·80	
78	Essex and Colchester	1862	90		275·				4·04	
79	Plymouth	1861–70	90	58·	4515·	10·	38·	3172	3·89	
80	Taunton	1861–70	96		806·				2·1	
81	Cheltenham Gen.	1861–70	84		581·				3·16	
82	Ipswich	1861–70			286·				4·23	
83	Bury St. Edmunds	1861–70	50	38·	805·	8·5	43·	7578	1·93	
83a	Jervis St., Dublin	1861–70	80		692·				4·37	
84	Mercer's, Dublin	1870	75		759·				5·00	
86	Stockport Inf.	1861–70	76		323·				2·95	
87	Preston Inf.	1861–70	70		167·				2·14	
88	Montrose Inf.	1861–70	70		269·				4·90	
91	Lancaster Inf.	1861–70	50		879·				9·87	
92	Arbroath Inf.	1861–70	33		167·				10·41	
93	Halifax Inf.	1861–70	60		309·				5·37	
94	Coventry & Warwickshire	1861–70	46	18·	254·	14·1	25·9	5388	3·81	
95	Dorchester	1865–70	50		415·				3·32	
96	Blackburn Inf.	1863–70	50	22·	202·	9·2	40·	7207	6·61	
97	Birkenhead Borough	1861–70	60		2515·				7·19	
98	Cardiff Inf.	1861–70	65		2505·				…	
99	Chichester Inf.	1861–70			306·				4·76	
99a	Guernsey Catel	1861–70			120·				12·87	

General Hospitals, 1861—70 (*continued*).

No.	Name of Hospital.	Years for which return is made.	Full No. of Beds.	Average Beds Occupied.	Average No. of In-patients.	Average No. of Patients to each Bed.	Mean Residence.	Mortality, per cent. of Beds.	Mortality, per cent. of Patients.	District Mortality per 1000.	Ratio of Hosp. to District Mortality.	Remarks.
100	Middlesborough Inf.	1865-70	60	...	286·	...	...	...	4·06	...	...	
101	Salford and Pendlebury	1861-70	60	...	156·5	...	...	...	6·26	...	...	
101a	West London	1866-70	25	...	215·	...	...	...	8·76	...	...	
104	Whitehaven Inf.	1861-70	56	...	209·	...	...	...	10·57	...	...	
105	Barrow-in-Furness	1866-70	18	...	67·5	...	...	...	3·73	...	...	
106	Swansea Inf.	1861-70	40	...	216·	·	...	...	4·20	...	...	
107	Surrey County	1866-70	54	...	268·	...	...	...	4·99	...	...	
107a	Roy. Cornwall Inf., Truro	1861-70	72	...	390·	...	...	...	1·84	...	...	
109	Aylesbury Inf.	1861-70	50	...	250·	...	...	...	3·11	...	...	
109a	Waterford City Inf.	1861-70	48	...	509·	...	...	...	3·10	...	...	
109b	St. Bartholomew's, Chatham	1863-70	68	...	405·	...	...	...	8·68	...	...	Exclusive of the ophthalmic beds.
109c	King's Lynn, Norfolk	1861-70	52	...	445·	...	...	...	3·28	...	...	
110	Poplar	1870	30	...	242·	...	...	...	10·33	...	...	
111	Durham County Inf.	1861-70	44	...	283·	...	...	...	...	...	...	Accident Hospital.
112	Fortar Inf.	1861-70	36	...	211·	...	...	...	...	...	...	
113	Huntingdon Inf.	...	42	...	...	...	...	...	4·08	...	...	
114	Peterborough Inf.	1861-70	42	...	121·	...	...	...	7·33	...	...	
115	Carnaardnen Inf.	1861-70	40	...	156·	...	...	...	2·76	...	...	
117	Chesterfield Inf.	1861-70	30	6·3	59·	9·36	39·	112·16	12·09	...	...	
118	Denbigh Inf.	1861-70	40	...	214·	...	...	...	3·17	...	...	
119	Hemel Hempstead, Herts	1861-70	40	...	249·	...	...	...	3·81	...	...	
120	Leith	1861-70	40	...	325·	...	...	...	8·42	...	...	
120a	Limerick City Inf.	1861-70	40	...	241·	...	...	...	4·36	...	...	
123	Metropolitan Free	1861-70	31	11·3	117·	12·2	28·	100·	6·43	...	...	
123a	Stamford and Rutland Inf.	1861-70	40	...	231·	...	...	...	3·90	...	...	
123b	Torbay Inf., Torquay	1861-70	40	...	163·	...	...	...	4·60	...	...	
123c	Tunbridge Wells Inf.	1851-70	36	...	130·	...	...	...	6·07	...	...	
124	Ayr Inf.	...	48	...	...	...	...	...	13·23	...	...	Accident Hospital.

No.	Hospital	Years	Stat. Soc. Journal.	Accident Hospital.	Accidents only.			Mortality %				Notes
126	Bridgewater Inf.	1861–70						...	194·		36	
127	Hertford Gen. Inf.	1861–70						2·09	143·		35	
129	Hitchin Inf.	1861–70						4·13	140·		32	
130	Hastings	1861–70						4·41	172·		32	
131	Belford, Inverness	1865–62		85·	26·	12·8		5·84	120·	10	30	
132	Great Northern, London	1861–62						6·66	128·		50	
133	Longton, Staffordshire	1869–70						7·61	52·		12	
134	Lowestoft	1861–70						5·37	76·		30	
135	North Ormsby, Yorks.	1861–70						4·71	144·		28	
136	Saffron Walden, Essex	1867–70						3·79	132·		30	
137	Stroud Gen.	1861–70						8·09	70·		20	
138	Great Yarmouth	1862–70						2·43	169·		28	
139	Ramsgate Seaman's	1861–70						4·07	64·		26	
139a	Bolton Inf.	1861–70						11·25	114·		40	
139b	Bootle Inf.	1869–70						4·27	58·		6	
141	Doncaster Inf.	1868–70						7·89	61·		16	
142	Seaham Harbour, Durham	1861–70						8·47	88·		24	
142a	Stratford-on-Avon Inf.	1861–70						3·08	99·		18	
145	Dover	1861–70						5·21	39·5			
146	Richmond, Surrey	1868–70						6·58	74·		15	
146a	Bangor Inf.	1861–70						2·98	48·		18	
147	Becket, Barnsley	1869–70						2·10	178·			
147a	Ditchingham, Norfolk	1870						5·61	140·		10	
148	French, London	1868–70						9·24	56·		20	
151	Sherborne, Dorset	1866–70						2·86	86·		20	
154	Sudbury, Suffolk	1868–70						1·16	69·		20	
155	Weston-super-Mare	1866–70						6·93	58·		15	
157	Darlington	1865–70						9·22	57·		18	
157a	Balfour, Kirkwall	1861–70						7·35	57·		12	
158	Clayton, Wakefield	1861–70						9·68	35·		12	
158a	Weybread, Suffolk	1868–70						3·81	65·		17	
159	Teignmouth Inf., Devon	1861–70						3·22				
160	Savernake Inf., Wilts	1866–70						2·93	83·		10	Entirely surgical.
160a	Loughborough	1865–70						4·24	53·		20	"Few accidents, and no incurables."
160b	Weymouth Inf.	1861–70						6·42	33·		10	
162	Festiniog	1866–70						8·	40·		15	

GENERAL HOSPITALS, 1861—70 (*continued*).

No.	Name of Hospital.	Years for which return is made.	Full No. of Beds.	Average Beds Occupied.	Average No. of In-patients.	Average No. of Patients to each Bed.	Mean Residence.	Mortality, per cent. of		District Mortality per 1000.	Ratio of Hosp. to District Mortality.	Remarks.
								Beds.	Patients.			
163	Gravesend	1861–70	15	...	70*	...	...	...	7·61	...	...	
165	Bodmin Inf., Cornwall	1861–70	14	...	46·5	...	...	...	1·72	...	...	
165a	Brecknock Inf.	1861–70	12	...	87·	...	...	...	4·25	...	...	
165b	Crewkerne, Somerset	1867–70	13	...	35·	...	...	...	5·71	...	...	
165c	Penrhyn, Wales	1866–70	12	...	66·	...	...	...	6·34	·	...	
168	Malvern Rural	1869–70	12	...	38·	...	...	...	2·63	...	...	
168a	Newport Inf.	1868–70	12	...	71·	...	...	...	4·69	...	...	
169	Tiverton	1868–70	12	...	18·	...	...	...	3·63	...	...	
171	Oswestry	1866–70	12	...	38·	...	...	...	7·89	...	...	
171a	Shepton-Mallet, Somerset	1869–70	12	...	5·5	...	...	...	...	...	...	
171b	Pembrokeshire Inf., Haverfordwest	1861–70	12	...	138·	...	...	...	1·66	...	...	
172	Reigate Inf., Surrey	1866–70	6	...	39·	...	...	...	3·57	...	...	
174	Bromley, Kent	1869–70	6	...	275	...	...	...	9·09	...	...	
175	Capel, Suffolk	1867–70	10	...	24·	...	...	...	5·37	...	...	
178	Ilfracombe, Devon	1865–70	10	...	27·	...	...	...	2·15	...	...	
178a	Tewkesbury	1865–70	7	...	56·	...	...	...	2·38	...	...	
178b	Wallasey, Cheshire	1867–70	4	...	22·	...	...	...	5·68	...	...	
180	Ashford, Kent	1870	9	...	49·	...	...	...	6·12	...	...	
182	Bridgenorth, Salop	1863–70	8	...	20·	...	...	...	3·12	...	...	
183	Cromer, Suffolk	1867–70	7	...	32·	...	...	...	2·36	...	...	
184	Dinorwic, Carnarvon	1867–70	8	...	36·	...	...	...	9·72	...	...	
184a	Fairford, Gloucester	1867–70	8	...	18·	...	...	...	3·07	...	...	
185	Fowey, Cornwall	1861–70	4	...	17·	...	...	...	1·76	...	...	
187	Bourton, Gloucester	1861–70	8	...	33·	...	...	...	2·14	...	...	
188	Mildenhall, Suffolk	1868–70	8	...	45·	...	...	...	3·59	...	...	
188a	Monmouth	1868–70	8	...	43·	...	...	...	3·84	...	...	
188b	Pembroke Inf.	1863–70	8	...	25·	...	...	...	4·54	...	...	
191	Shedfield, Hants	1867–70	8	...	18·5	...	...	...	2·70	...	...	

										Accident Hospital
191a	Tetbury	1868-70	3		24·			1·37		
193	Stockton-on-Tees	1863-70	6		35·			12·87		
193a	Alton, Hants	1869-70	7		31·				: o	
195	Burford, Salop	1870	7		33·			0·		
196	Chesham, Bucks	1869-70	7		28·			3·57		
197	Dunster, Somerset	1867-70	5		20·			6·66		
197a	East Grinstead, Sussex	1864-70	7		29·5			3·36		
199	Mansfield Woodhouse, Notts	1867-70	…		40·			1·87		
203	Royston, Herts	1870	6		19·			5·26		
204	St. Albans, Herts	1870	5		12·			0·	: o	
208	Burford, Oxon	1869-70	6		20·			5·		
210	Clearwell, Gloucester	1869-70	6		12·			6·25		
210a	Cranleigh, Suffolk	1861-70	6		19·			5·72		
210b	Downton, Wilts	1870	6		27·			3·70		
210c	Driffield, Yorks	1867-70	5		31·			3·2		
213	Guisborough, Yorks	1865-70	6		14·			5·81		
213b	Hambrook, Gloucester	1866-70	6		46·			3·29		
213b	Harrogate Cottage	1870	5		20·			0·	o	
214	Iver, Bucks	1863-70	3		15·			4·63		
214a	King's Sutton, North Hants	1868-70	6		32·			4·16		
217	Warminster, Wilts	1867-70	7		22·5			3·37		
218	Newick, Suffolk	1869-70	6		21·			3·70		
219	Petworth, Sussex	1867-70	5		20·			7·69		
221	Speen, Bucks	1869-70	6		30·			3·33		
222	Wirksworth, Derby	1867-70	4		17·5			2·87		
224	Buckhurst Hill, Essex	1867-70	5		28·			2·65		
225	Bronyard, Hereford	1869-70	6		30·			1·75	o	
227	Melksham, Wilts	1869-70	4		31·5			0·		
227a	Worksop, Notts	1870	4		20·			10·	o	
230	Yate, Gloucester	1870	5		14·			0·		
230a	Crimond, Aberdeen	1865-70	4		23·			2·60		
230b	Charmouth, Dorset	1867-70	3		12·			4·16		
235	Market Rasen, Lincoln	1869-70	2		10·5			14·28		

IRISH COUNTY INFIRMARIES, 1861—70.

No.	Name of Hospital.	Years for which return is made.	Full No. of Beds.	Average Beds Occupied.	Average No. of In-patients.	Average No. of Patients to each Bed.	Mean Residence.	Mortality, per cent. of		District Mortality per 1000	Ratio of Hosp. to District Mortality.	Remarks.
								Beds.	Patients.			
3	Maryborough	1861–70	100	43·6	617·	14·1	26·	41·97	2·96	...	...	
4	Roscommon	1861–70	85	...	708·	...	...	...	2·17	...	...	
5	Tyrone	1861–70	65	65	679·	10·44	35·	16·00	1·53	...	...	
7	Down Patrick	1861–70	52	...	422·	...	...	...	4·00	...	...	
8	Armagh	1861–70	72	...	1178·	...	...	...	1·72	...	...	
10	Wexford	1861–70	72	36·3	437·	12·	30·4	35·53	2·95	...	...	
11	Cavan	1863–70	70	50	660·	13·2	29·1	12·44	1·06	...	...	
12	Donegal	1861–70	80	46·3	529·	11·2	32·5	25·48	2·23	...	...	
14	Clare	1861–70	60	64	924·	14·4	25·3	21·56	1·49	...	...	
15	Mayo	1861–70	70	60	700·	11·66	31·4	18·83	1·61	...	...	
18	Fermanagh	1861–70	60	55	748·	13·6	26·8	30·54	2·24	...	...	
20	Leitrim	...	52	...	362·	...	...	...	2·12	...	...	
21	King's County	1861–70	50	33	374·	11·3	32·3	22·72	2·00	...	...	
22	Louth	1861–70	45	...	402·	...	...	...	2·19	...	...	
23	Carlow	1861–70	40	...	258·	...	...	...	2·79	...	...	
24	Kerry	1861–70	50	42	518·	12·3	29·7	17·14	1·39	...	...	
27	Longford	1861–70	39	...	352·	...	...	...	1·84	...	...	

FEVER HOSPITALS, 1861—70.

No.	Name of Hospital.	Years for which return is made.	Full No. of Beds.	Average Beds Occupied.	Average No. of In-patients.	Average No. of Patients to each Bed.	Mean Residence.	Mortality per cent. of Beds.	Mortality per cent. of Patients.	District Mortality per 1000.	Ratio of Hosp. to District Mortality.	Remarks.
	London Fever	1861–70	320	...	2911·	...	...	...	16·88	...	...	
	Fever, Cork	1861–70	160	...	1075·	...	...	...	7·06	...	...	
	Cork Street, Dublin	1861–70	120	81·7	1593·	19·5	19·	142·10	7·29	...	...	Statistics doubtful.
	Limerick, St. John's	1861–70	100	27·	251·	9·3	39·2	52·96	5·70	...	...	
	Newcastle-upon-Tyne	1861–70	45	...	260·	...	...	...	16·49	...	...	
	Bedford	1861–70	50	...	75·4	...	...	...	11·40	...	...	
	County Down	1861–70	40	...	42·	...	...	...	10·28	...	...	
	Queenstown, Cork	1861–70	40	...	128·	...	...	...	8·99	...	...	
	Gateshead	1866–70	36	...	96·	...	...	...	11·06	...	...	
	Ayr	1861–70	48	...	80·	...	...	...	13·23	...	...	
	Monaghan	1861–70	39	...	23·	...	...	...	9·17	...	...	
	Carlisle	1861–70	30	...	52·	...	...	...	16·60	...	...	

CHILDREN'S HOSPITALS, 1861—70.

No.	Name of Hospital.	Years for which return is made.	Full No. of Beds.	Average Beds Occupied.	Average No. of In-patients.	Average No. of Patients to each Bed.	Mean Residence.	Mortality per cent. of Beds.	Mortality per cent. of Patients.	District Mortality per 1000.	Ratio of Hosp. to District Mortality.	Remarks.
	Ormond Street, London	1861–70	127	65·8	645·	9·8	37·2	112·00	11·42	...	...	
	Evelina, London	1869–70	100	...	185·	...	...	...	4·33	...	...	
	Liverpool, Myrtle Street	1861–70	60	...	126·	...	...	...	7·30	...	...	
	Edinburgh	1861–70	70	44·	412·	9·36	39·	96·59	10·34	...	...	
	Birmingham	1863–70	55	21·	508·	24·2	11·	177·9	7·35	...	...	
	Victoria, London	1870	40	...	251·	...	...	...	6·37	...	...	
	Bristol	1867–70	40	27·	348·	12·9	28·2	38·53	3·01	...	...	
	Brighton	1869–70	10	...	49·	...	...	...	4·08	...	...	
	Gloucester	1867–70	24	...	17·5	...	...	...	2·55	...	...	
	North Eastern, London	1870	12	...	98·	...	...	...	9·17	...	...	
	Jenny Lind, Norwich	1861–70	20	...	63·4	...	...	...	4·57	...	...	
	Belgrave, London	1867–70	16	...	85·5	...	...	...	4·38	...	...	

GENERAL HOSPITALS, 1870—75.

* Hospitals from which published Reports have been received.

No.	Name of Hospital.	Years for which return is made.	Full No. of Beds.	Average Beds Occupied.	Average No. of In-patients.	Average No. of Patients to each Bed.	Mean Residence.	Mortality, per cent. of Beds.	Mortality, per cent. of Patients.	District Mortality per 1000.	Ratio of Hosp. to District Mortality.	Remarks.
1	St. Bartholomew's	1870-5	710	301·49	3337·16	11·11	32·85	56·91	5·12	26	1·972	Excluding ophthalmic beds and cases.
2*	Guy's	1870-5	650	539·6	5617·	10·4	35·	99·	9·5	25	3·8	
3*	London	1870-5	600	506·6	5240·	10·34	34·42	121·98	11·8	30	3·93	12·8 of all cases are fever. *See* Fever Hosps. 1·2 of all deaths.
4*	Roy. Inf., Glasgow	1870-5	584	475·3	5591·	11·76	30·6	126·44	11·1	...	...	Reports very complete.
5	St. Thomas's	1871-5	572	347·8	3142·7	9·03	36·35	109·54	12·13	25	4·852	
6*	Roy. Inf., Edin.	1870-5	565	411·5	4525·3	10·99	33·01	111·2	10·18	...	...	Of all cases { 5% Fever 40% Surgical 55% Medical
7*	St. George's	1872 & 5	353	338·6	3933·5	11·6	31·46	85·65	7·37	19	3·88	Sq. space per bed, 106 feet. Cubic space, do., 2099·5 ft.
8*	Leeds Gen. Inf.	1870-5	310	200·	2943·16	14·21	25·62	100·	6·78	30	2·26	
9*	Middlesex	1870-5	305	219·93	2041·3	9·28	39·3	119·54	12·92	25	5·168	Three wards retained for cancer cases. Reports very incomplete.
10	Roy. Inf., Aberdeen	1870-5	300	127·8	1720·8	13·46	29·37	89·	6·61	...	...	Admits fever and smallpox. *See* Fever Hosps. 28·4% of all the cases are fever.
11*	Manchester Inf.	1870-5	296	205·16	2901·3	14·14	25·8	152·46	10·78	32	3·37	
12*	Dundee Inf.	1870-5	280	131·5	1692·5	12·8	30·12	123·57	9·6	...	...	Reports very complete.
13*	Liverpool Inf.	1870-5	270	227·16	2555·3	11·24	32·4	81·35	7·23	39	2·5	Lock and asylum cases excluded.
14*	Birmingham Gen.	1870-5	256	207·3	2713·3	13·08	27·8	105·72	8·07	27	3·	Reports incomplete.

No.	Hospital	Years	Beds									Remarks
15*	Steevens's, Dublin	1870-5	250	142·82	2000·3	14·	24·63	39·21	2·8	...	...	From Parliamentary Repts. No information obtained.
16	Bristol Roy. Inf.	1870-5	242	...	...	...	...	...	...	25	1·364	
17*	Devon and Exeter	1871-5	230	175·5	1361·16	7·81	47·8	28·37	3·41	...	...	Reports very incomplete.
18*	Mater Miser., Dub.	1870-5	230	137·2	2057·3	15·8	19·77	111·88	7·46	28	2·918	
19*	Newcastle Inf.	1870-5	230	156·6	1661·	10·6	34·43	86·71	8·17	...	...	Exclusive of Lock beds and cases.
20*	Royal Albert, Devonport	1870-5	218	46·54	378·86	8·13	50·33	43·65	5·36	26	...	
21*	Leicester Inf.	1870-5	210	158·16	1955·	15·	29·66	57·15	4·62	24	1·77	Fever { 5% of all cases. { 16% of all deaths.
22*	Wolverhampton Gen.	1871-5	210	...	...	...	...	...	6·78	...	2·86	Great preponderance of surgical cases, especially accidents.
23*	Liverpool, South {Old {New	1870-2 / 1873-6	120	89·3 / 124·3	1449·3 / 2089·6	16·2 / 16·81	22·5 / 21·71	122·73 / 97·58	7·57 / 5·81	39	1·5	
24*	Glasgow Western	1875	200	139·	1253·	9·	36·	81·29	9·02	27	3·41	Opened in 1874. Report full.
25*	Westminster	1870-5	200	143·	1813·	12·6	28·9	116·83	9·21	27	3·41	
26*	Whitworth, Hardwicke & Richmond, Dub.	1870-5	192	183·56	2811·	15·8	22·29	112·22	7·7	...	...	Including fever cases in Hardwicke Hosp, which were 26·7% of the whole. See Fever Hosps.
27*	Royal Portsmouth	1870-5	180	47·16	361·6	7·66	47·6	52·58	6·85	21	3·36	Lock cases & beds excluded.
28*	North Stafford Inf.	1870-5	177	122·27	1300·	10·66	33·57	59·3	5·57	22	2·533	
29*	Derby Gen. Inf.	1870-5	175	161·6	1092·8	10·75	34·	86·41	8·04	23	35·	
30*	Paisley Inf.	1870-5	174	57·37 / 80·7	745·8 / 1096·6	13· / 13·58	28·07 / 26·12	100·8 / 108·55	7·7 / 7·98	...	...	Without fever. / With fever, of which there was 28% of all cases.
31*	Sussex County	1874-5	170	128·	1237·5	9·6	38·	76·56	7·93	22	3·6	Admirable reports.
32*	King's College	1870-5	170	142·24	1685·5	11·84	30·8	143·41	12·5	25	4·82	Reports very incomplete.
33	Radcliffe Inf., Oxford	1870-5	161	136·	1237·	9·01	40·5	38·84	4·27	22	1·94	Reports incomplete.
34*	St. Mary's, Paddington	1873-4	170	146·2	1812·	12·3	29·44	141·23	11·36	21	5·41	
35*	Belfast Gen.	1870-5	160	111·3	1699·8	15·27	23·9	104·67	6·85	...	...	
36*	Sheffield Gen. Inf.	1871-5	160	136·5	1161·25	8·5	34·43	93·59	11·	29	3·8	
37*	Bristol Gen.	1871-5	150	108·	1404·	12·92	28·25	85·91	6·6	29	2·276	Reports very full.
38*	Charing Cross	1870-5	150	119·1	1277·6	10·72	34·	124·47	11·6	25	4·64	
39	Chester Gen. Inf.	1870-5	150	57·46	700·5	12·19	30·	77·09	6·32	22	2·872	
40*	Queen's, Birmingham	1870-5	150	129·	1596·3	12·37	29·5	95·	7·67	27	2·824	
41*	University Coll., Lond.	1870-5	150	133·58	1703·5	12·75	29·4	152·34	11·91	23	5·178	Admirable reports.

GENERAL HOSPITALS, 1870—75 (*continued*).

No.	Name of Hospital.	Years for which return is made.	Full No. of Beds.	Average Beds Occupied.	Average No. of In-patients.	Average No. of Patients to each bed.	Mean Residence.	Mortality, per cent. of		District Mortality per 1000.	Ratio of Hosp. to District Mortality.	Remarks.
								Beds.	Patients.			
42*	Hull Gen. Inf.	1870-5	150	94·08	1094·5	11·6	31·6	103·72	8·93	26	3·434	
43*	Liverpool Northern	1870-5	144	116·9	1613·2	13·8	26·44	88·62	6·42	39	1·646	Reports incomplete.
44*	Nottingham Gen.	1870-5	142	...	1105·	...	...	...	7·87	24	3·28	
45*	Bradford Gen. Inf.	1870-5	140	88·55	818·3	9·24	39·5	72·61	7·86	25	3·144	Reports full.
46*	Gloucester Gen. Inf.	1870-5	140	86·69	819·2	9·33	39·12	50·52	5·34	21	2·543	Without fever cases.
47*	Greenock Gen. Inf.	1870-5	140	{46·53 / 79·2}	{543· / 1094·6}	{11·67 / 13·87}	{31·25 / 26·31}	{146·57 / 169·19}	{12·56 / 12·24}	...	...	With fever, which is almost 50 % of all the cases. Admirable reports.
48	Royal Berks	1870-5	140	107·6	1025·	9·6	41·6	39·59	4·13	22	1·877	
49*	Salop Inf	1871-5	140	98·3	923·2	9·4	38·68	60·22	6·4	25	2·76	Reports defective.
50*	Northampton Gen. Inf.	1870-5	138	119·	1182·4	9·85	39·	42·52	4·28	24	1·783	
51*	City of Dublin	...	130	...	...	...	...	...	...	...	...	No details in reports.
52*	Addenbrooke's, Cambridge	1870-5	120	83·8	799·6	9·54	38·26	55·48	5·81	22	2·64	
53*	Huddersfield Inf.	1870-5	120	56·5	485·	8·58	39·5	64·42	7·58	24	3·175	Reports very defective.
54*	Meath, Dublin	1870-5	120	89·64	1174·8	13·2	26·18	84·78	6·5	...	...	Parliamentary reports.
55*	Royal Bath	1870-5	120	76·5	1013·	13·24	28·3	118·17	8·04	22	3·654	
56*	Norwich and Norfolk	1870-5	118	103·6	848·	8·18	42·3	62·74	77	24	3·208	Reports full. Mort. for 90 years, 5·5.
57*	Perth Inf.	1874-5	110	...	532·	...	...	...	7·51	...	...	Admits fever.
58*	Sunderland Gen. Inf.	1872-5	110	53·	497·75	9·39	38·8	72·92	7·8	24	3·25	Admits fever.
59	Hants County	1870-5	108	75·	707·3	9·43	38·7	40·66	4·35	19	2·29	
60*	Kent and Canterbury	1870-5	104	67·	521·	7·88	46·83	49·49	6·37	24	2·655	
61*	Dumfries Inf.	1870-5	104	32·81	370·3	11·28	28·35	76·19	6·75	...	...	Admirable reports.
62*	Sheffield Public	1874-5	104	...	767·5	...	...	...	8·14	29	2·807	Reports extremely defective.
63*	Adelaide, Dublin	1870-5	100	...	847·	...	...	...	7·14	...	...	Admits fever. Reports defective.
64*	North Devon, Barnstaple	1870-5	100	62·3	593·6	9·53	38·25	17·9	1·88	18	1·047	
65*	Bedford Gen. Inf.	1870-5	100	...	680·8	...	...	...	3·65	20	1·825	Reports incomplete.

No.	Hospital	Years										Remarks
66*	Carlisle Inf.	1870-5	100	66·66	4195·	...	...	...	6·03	24	2·512	Returns incomplete.
67	Cork, North, Charitable	1870-5	100	59·	854·16	12·83	28·4	67·74	5·26	...	...	Admits fever, and cases of phthisis are about 10% of whole, it being an hospital chiefly for Germans.
68	German, Dalston	1870-5	100	...	919·	15·58	23·49	54·23	10·1	20	5·05	
69*	Leamington	1874-5	100	...	508·	...	...	...	4·	20	2·	An absolutely free hospital.
70	Lincoln County	1870-5	100	77·	727·	...	38·76	37·66	4·	20	...	
71*	Royal Free, London	1870-5	100	...	1283·4	9·3	...	...	6·96	23	3·03	
72*	Salisbury Gen. Inf.	1870-5	100	80·48	912·2	11·33	31·08	41·5	3·66	20	1·83	
73*	Southampton Inf.	1870-5	100	81·27	846·	10·4	35·	61·	6·	23	2·609	
74*	Worcester Inf.	1870-5	100	71·35	800·3	11·21	32·56	59·08	5·26	25	2·104	
75	St. Vincent's, Dublin		100	...	...	...	...	...	...	...	...	Information could not be given.
76*	York County	1870-5	100	72·9	721·5	9·9	36·8	51·57	5·2	23	2·26	
77*	Stafford Gen. Inf.	1870-5	96	49·5	6786·	13·7	23·8	54·14	3·95	20	1·975	Admits fever. Full reports
78	Es-ex and Colchester	1870-5	94	43·5	301·	6·9	52·5	29·9	4·	22	1·818	No information obtained.
79	Plymouth		94	...	...	...	...	...	...	23	...	
80*	Taunton	1870-5	92	80·3	548·	6·77	53·8	36·31	5·32	23	2·8	
81*	Cheltenham Gen.	1870-5	90	60·3	571·16	9·47	38·6	31·77	3·35	19	1·763	
82	Ipswich		90	...	...	...	...	...	...	22	...	No reply.
83*	Bury St. Edmunds	1870-5	84	...	422·	...	...	...	3·57	23	1·552	Reports very incomplete.
84*	Mercer's, Dublin	1870-5	80	...	928·	...	...	...	6·	...	...	Reports very defective.
85	Sir P. Dun's, Dublin	1870-5	80	...	...	...	...	...	...	...	...	Information promised.
86*	Stockport Inf.	1870-5	80	41·16	378·5	11·44	32·5	83·33	9·04	25	3·616	Reports defective.
87*	Preston Inf.	1870-5	76	...	469·	...	...	...	7·3	28	2·67	
88*	Montrose Inf.	1874-5	76	...	202·8	...	...	...	6·82	...	...	Smallpox & fever admitted.
89*	Hereford Inf.	1870-5	74	...	626·	...	...	...	5·5	21	2·619	Defective returns.
90*	Guest, Dudley	1872-5	73	...	342·	8·5	43·	84·1	7·46	25	2·984	
91*	Lancaster Inf.	1870-5	70	13·	110·8	...	...	...	9·9	20	4·95	Reports very defective.
92*	Arbroath Inf.	1870-5	70	...	144·16	...	...	...	11·	...	...	Admits fever. Reports defective.
93	Halifax Inf.	1870-5	70	...	...	...	...	...	...	24	...	"Records inaccurate and of no value."
94*	Coventry & Warwicksh.	1870-5	62	21·16	234·8	10·62	34·3	68·05	6·09	21	2·9	Reports full.
95*	Dorchester	1870-5	60	53·5	3915·	7·3	50·	32·33	4·42	19	2·326	
96*	Blackburn Inf.	1870-5	60	36·	486·	13·5	28·6	81·94	5·95	26	2·288	

GENERAL HOSPITALS, 1870—75 (continued).

No.	Name of Hospital.	Years for which return is made.	Full No. of Beds.	Average Beds Occupied.	Average No. of In-patients.	Average No. of Patients to each Bed.	Mean Residence.	Mortality, per cent. of		District Mortality per 1000.	Ratio of Hosp. to District Mortality.	Remarks.
								Beds.	Patients.			
97*	Birkenhead Borough	1871-5	60	...	472·4	...	...	...	4·	23	1·739	Reports defective.
98	Cardiff Inf.	1870-5	60	40·4	418·	10·34	35·3	77·1	7·46	21	3·561	
99	Chichester Inf.	1870-5	60	32·3	276·5	8·56	38·	40·24	4·85	19	2·552	
100*	Middlesborough Inf.	1870-5	60	40·4	419·3	10·37	35·2	69·3	6·7	...	...	
101*	Salford and Pendleton	1875	60	...	438·	...	...	...	8·7	27	3·222	Report defective.
102*	Wigan Inf.	1873-5	60	30·3	215·5	7·1	38·39	62·7	8·8	29	3·034	Large proportion of "smash" accidents.
104	Whitehaven Inf.	1870-5	56	15·3	217·	14·16	25·76	23·6	14·36	25	5·744	Accidents only.
105	Barrow-in-Furness	1870-5	55	9·16	175·	19·1	9·	13·74	7·	21	3·333	
106*	Swansea Inf.	1870-5	55	4·15	312·8	8·26	44·18	46·5	6·17	22	2·804	
107	Surrey County	1870-5	54	38·8	374·	9·62	37·75	51·04	5·34	19	2·81	
108	Ardwick	...	50	...	...	...	...	...	...	...	...	Not yet in existence, though returned in "Med. Direct."
109*	Aylesbury Inf.	1870-5	50	23·55	369·	9·21	39·63	36·42	3·95	21	1·88	Entirely for accidents. Information promised.
110	Poplar	...	48	...	...	...	...	...	...	...	...	
111	Durham County Inf.	1870-5	44	25·2	220·	8·7	42·	58·	6·06	21	2·885	Admits fever.
112	Forfar Inf.	1871-5	44	64·	686·	10·72	34·	100·	9·3	...	...	
113	Huntingdon	1870-5	42	26·47	236·	8·91	40·96	34·6	3·88	...	...	
114	Peterborough	1870-5	42	10·08	127·	12·6	29·	115·07	9·17	20	4·585	
115	Carmarthen	1870-5	40	20·	167·	8·4	43·5	30·	3·6	20	1·8	
116*	Chalmers', Edin.	1870-5	40	27·62	224·	8·11	45·	63·36	7·81	...	...	For accidents.
117	Chesterfield Inf.	1870-5	40	13·55	126·5	9·33	37·4	103·32	11·11	22	5·05	"No records kept."
118	Denbigh	...	40	...	...	...	...	...	...	...	...	
119	Heml. Hempstead, Hts.	1870-5	40	28·6	302·16	10·56	34·5	32·02	3·33	20	1·665	
120	Leith	...	40	...	...	...	...	...	...	...	...	"Statistics not available."
121	Macclesfield Gen. Inf.	1873-5	40	29·6	228·	7·7	47·	68·27	8·	23	3·478	Opened 1873.
122	Maidstone Gen.	...	40	...	...	...	...	...	...	21	...	Information promised.
123	Metropolitan Free	...	40	...	...	...	...	...	...	26	...	Information promised.

		1875	38	14'	190'	13'5	28'	50'	7'4 / 11'			
124*	Ayr Inf.	...	38							...	...	Without fever. All cases, 38 % being typhus. *See* Fever Hosps.
125*	Burton-on-Trent	1870-5	37	12'	142'25	11'85	30'8	98'94	10'54	21	5'019	Each in-ptnt pays 2s. a wk.
126*	Bridgewater	1870-5	36	15'16	289'6	19'1	19'	50'72	3'	19	...	
127	Hertford Gen. Inf.	1870-5	35	18'3	170'5	9'32	45'5	235'	3'8	19	2'	
128*	Oldham	1872-5	33	17'5	166'	9'5	39'1	134'85	14'3	26	5'5	
129	Hitchin	...	32	14'3	158'8	11'11	33'16	32'16	2'9	20	1'45	
130	Hastings	1870-5	32	...	250'	...	...	...	2'92	20	1'46	
131	Belford, Inverness	1870-5	30	11'2	127'	11'33	27'08	17'85	1'57	...	...	
132	Gt. Northern, London	1870-5	30	21'3	259'5	12'17	29'97	115'02	6'3	21	3'	
133	Longton, Staffordshire	1870-5	30	...	108'	...	...	...	4'8	...	...	
134	Lowestoft	1870-5	30	16'39	130'16	7'94	45'95	43'68	5'5	18	3'055	Books stated to give no information.
135	North Ormsby, Yorks	...	30	...	...	...	...	...	...	...	...	
136*	Saffron Walden, Essex	1871-5	30	15'75	170'4	10'8	31'94	54'6	5'05	19	2'057	
137	Stroud Gen.	1870-5	30	...	117'	...	35'2	49'	6'5	21	3'095	Accident Hospital.
138	Great Yarmouth	1870-5	28	15'3	158'6	10'36	30'6	38'1	4'73	24	1'97	
139*	Ramsgate, Seaman's	1870-5	27	10'05	118'6	11'8	30'6	93'02	3'2	...	...	Reports very complete.
140*	Rotherham	1873-5	25	12'9	140'3	10'8	30'6	93'02	8'54	24	3'558	
141	Doncaster Inf.	1870-5	24	11'16	110'16	9'87	34'54	91'04	9'22	...	...	
142	Seaham Harbour, Durham		24	2'	411'6	20'58	18'5	14'5	6'9	22	3'136	
143	Walsall	1870-5	24	16'92	194'6	11'5	31'73	93'38	8'13	24	3'387	
144	West Bromwich	1870-5	24	13'25	156'75	11'9	25'75	81'13	6'86	23	3'119	
145	Dover	1870-5	21	13'95	138'5	9'92	30'06	53'76	5'42	20	2'71	
146	Richmond, Surrey	1870-5	21	13'4	151'6	11'3	32'3	89'45	8'13	19	4'279	
147	Becket, Barnsley	1872-5	20	6'03	71'	11'77	31'	82'91	7'04	25	2'813	
148	French, London	1870-5	20	10'5	146'6	14'	21'	120'	8'6	23	3'739	Reports defective.
149°	Stanley, Liverpool	...	20	...	...	...	...	...	...	39	...	
150	Louth, Lincoln	1873-5	20	6'67	65'6	9'83	37'13	30'	3'05	19	1'605	
151*	Sherborne, Dorset	1870-5	20	12'5	120'6	9'6	41'5	36'	3'73	19	1'963	Admits fever. Reports defective.
152*	Southport Inf.	1870-5	20	...	190'3	...	...	...	8'4	...	...	
153*	South Shields Inf.	1873-5	20	10'3	110'3	10'7	28'3	97'	9'07	20	9	
154*	Sudbury, Suffolk	1870-5	20	...	109'4	...	...	1'8	1'8		2'588	
155	Weston-super-Mare	1870-5	20	10'06	121'5	12'97	28'13	59'64	4'93	19		

GENERAL HOSPITALS, 1870—75 (*continued*).

No.	Name of Hospital.	Years for which return is made.	Full No. of Beds.	Average Beds Occupied.	Average No. of In-patients.	Average No. of Patients to each Bed.	Mean Residence.	Mortality, per cent. of Beds.	Mortality, per cent. of Patients.	District Mortality per 1000.	Ratio of Hosp. to District Mortality.	Remarks.
156	Yeovil, Somerset	1874–5	20	2·6	29·	11·23	32·5	57·69	5·18	19	2·726	
157	Darlington	1870–5	19	7·7	74·3	9·6	38·	62·72	6·53	21	3·109	
158	Clayton, Wakefield	1870–5	18		89·6			34·07	8·14	23	3·537	Accident Hospital.
159	Teignmouth, Devon	1870–5	17	6·75	77·6	1·15	31·74	34·07	3·33	...	...	
160	Savernake, Wilts	1870–5	16	12·66	138·3	10·92	33·16	34·43	3·	...	...	
161	Banbury, Oxfordshire	1872–5	15	8·75	112·	12·8	21·	51·51	3·7	20	1·85	
162	Festiniog	1870–5	15	3·3	37·8	11·4	32·	106·	9·2	22	4·18	Accident Hospital.
163	Gravesend	1870–5	15	4·7	84·6	18·	20·5	191·5	10·63	22	4·83	
164	Rugeley	1872–5	15	6·26	59·5	9·5	38·25	67·89	7·14	...	...	
165	Bodmin, Cornwall	1870–5	14	4·2	50·6	12·	30·	11·9	·99	18	·55	
166	Med. Miss., Edin.		13										No details available. Hosp. closed on account of fever.
167	Chelmsford	1870–5	12	1·34	35·	26·	14·	120·	4·5	19	2·368	
168	Malvern, Rural	1870–5	12	3·6	42·5	11·86	36·5	55·55	4·7	20	...	Accident Hospital.
169	Tiverton	1870–5	12	1·72	24·16	14·	26·	87·	6·2	20	3·1	
170	Beccles, Suffolk	1874–5	12		33·				10·97	...	...	Opened 1874.
171	Oswestry	1870–5	12	4·5	56·6	12·5	29·2	85·11	6·7	21	3·19	
172	Reigate, Surrey	1870–5	12	8·28	80·6	9·73	37·5	38·16	4·	...	...	
173	Jarrow-upon-Tyne	1870–5	11	5·4	81·	15·	24·3	70·37	4·7	...	...	Accident Hospital.
174	Bromley, Kent	1870–5	10		31·				2·16	...	...	
175	Capel, Suffolk	1871–5	10		49·75				2·	...	...	
176*	Ealing	1875	10	2·21	31·	14·	26·	45·45	3·22	20	1·61	
177*	Frome, Somerset		10										No information in report.
178*	Ilfracombe, Devon		10										
179	Newton, Devon	1874–5	10	4·	44·5	11·12	44·5	75·	6·74	...	...	
180	Ashford, Kent	1870–5	9		62·				7·7	...	...	
181	Epsom	1873–5	9	3·25	48·6	14·8	24·6	70·8	4·8	18	...	
182	Bridgnorth, Salop		8										"Books so loosely kept that it is impossible to make any return."

No.	Place	Years	n	a	b	c	d	e	f	g	h	Remarks
183	Cromer, Suffolk	1870-5	8	…	…	32.8	…	…	3.45	…	…	No statistics kept.
184	Dinorwic, Carnarvon	1870-5	8	…	…	13.3	…	…	3.76	…	…	No statistics kept.
185	Fowey, Cornwall	…	8	…	…	44.	…	…	…	…	…	Information promised.
186	Hatfield, Essex	…	8	…	…	51.8	…	…	3.76	…	…	Information promised.
187	Bourton, Gloucester	1870-5	8	3.75	11.2	44.	30.96	13.33	1.13	19	1.826	Circulars returned blank; no reports to be had.
188	Millenhall, Suffolk	1871-5	8	5.3	9.76	51.8	37.4	…	3.47	21	…	
189	Luton, Beds.	…	8	…	…	…	…	…	…	…	…	
190	Seacombe, Cheshire	1872-5	8	4.5	11.77	53.	28.25	16.66	9.91	…	…	
191	Shelfield, Hants	1871-5	8	1.75	16.1	28.25	29.	71.42	4.42	…	…	
192*	Rochdale	1872-5	8	…	…	41.	…	…	12.8	24	5.33	Accident Hospital. Statistics badly kept.
193	Stockton-on-Tees	1870-5	8	3.13	15.2	47.6	24.	64.85	10.84	23	4.713	
194	Wrexham	…	7	…	…	74.3	…	…	…	…	…	
195	Burford, Salop	1870-5	7	2.83	10.4	29.5	35.	45.93	4.34	…	…	
196*	Chesham, Bucks	1870-5	7	3.11	7.41	23.	49.25	31.26	4.35	…	…	
197	Dunster, Somerset	1870-5	7	5.	6.36	31.8	23.2	6.6	1.03	…	…	Admirable reports: many important operations detailed. Patients pay a small sum.
198*	Erith, Kent	1872-5	7	4.37	15.7	68.75	…	108.69	7.01	…	…	
199	Mansfield Woodhouse, Notts	1870-5	7	3.	16.2	48.	22.5	50.	3.12	22	1.418	
200	Moreton, Gloucester	1873-5	7	2.15	12.5	31.6	29.1	80.	5.6	…	…	
201	New Swindon, Wilts	1872-5	7	2.17	13.	28.15	28.	57.6	4.	…	…	
202	Petersfield, Hants	1871-5	7	3.43	10.31	35.4	31.2	75.8	7.35	…	…	
203*	Royston, Herts	1870-5	7	3.88	9.15	35.5	38.4	55.68	6.15	18	…	
204	St. Albans, Herts	1870-5	7	2.8	15.5	44.3	23.5	35.71	2.25	19	…	
205	Otery St. Mary, Devon	1870-5	7	3.8	11.84	45.	19.8	31.58	2.66	19	…	"No statistics of any use."
206	Tenby, Pembrokeshire	…	6	3.2	8.75	28.	42.	31.25	3.57	…	3.416	
207	Bovey Tracey, Devon	1872-5	6	2.6	11.5	30.	31.6	70.38	6.1	…	1.4	
208	Burford, Oxon	1870-5	6	3.38	10.52	35.5	42.	73.9	.7	…	…	
209	Chalfont S. Peter, Bucks	1871-4	6	1.03	8.7	9.	42.	24.27	2.77	…	…	
210	Clearwell, Gloucester	1872-5	6	3.	8.3	25.	31.2	26.6	3.2	20	…	
211	Dawlish, Devon	1871-5	6	2.	9.5	18.5	38.1	62.5	6.76	19	1.6	
212	Devizes, Wilts	1872-5	6	2.86	8.44	24.1	36.1	46.5	5.5	22	3.557	
213	Guisborough, Yorks	1870-5	6	1.56	9.7	15.2	37.5	38.46	4.	18	2.5	
214*	Iver, Bucks	1870-5	6	.5	7.5	37.5	54.	150.	13.3	…	…	Admirable reports.
215	Launceston, Cornwall	1872-5	6	…	…	3.75	…	…	…	…	7.388	Admirable reports. Accidents on railway.

GENERAL HOSPITALS, 1870—75 (*continued*).

No.	Name of Hospital.	Years for which return is made.	Full No. of Beds.	Average Beds Occupied.	Average No. of In-patients.	Average No. of Patients to each Bed.	Mean Residence.	Mortality, per cent. of		District Mortality per 1000.	Ratio of Hosp. to District Mortality.	Remarks.
								Beds.	Patients.			
216	Milton Abbas, Dorset	1874-5	6	3·	22·5	7·4	49·	0·	0·	..	..	
217	Warminster, Wilts	1870-5	6	3·6	37·	10·13	36·	14·8	4·06	18	2·255	
218	Newick, Suffolk	1870-5	6	4·3	20·6	4·8	75·6	23·27	4·83	..	..	Exclusive of fever cases.
219	Petworth, Sussex	1873-5	6	3·6	30·	8·33	42·	27·77	3·33	24	2·416	Accident Hospital.
220	Ruabon	1870-5	6	2·55	27·	10·58	34·5	13934	5·8	..	..	
221	Speen, Bucks	1870-5	6	4·05	38·6	9·53	39·55	13·82	1·67	..	..	
222*	Wirksworth, Derby	1870-5	6	3·01	31·2	10·34	35·3	33·22	3·2	..	..	Average age of patients, 29·5 years.
223	Worthing, Sussex	..	6	..	..	..	..	..	..	..	..	Only a dispensary.
224	Buckhurst Hill, Essex	1870-5	5	4·4	33·	10·14	36·	7·5	3·	..	..	
225	Bromyard, Hereford	1870-5	5	1·4	17·	12·3	29·6	71·43	·92	19	·488	
226*	Hayes, Middlesex	1874-5	5	2·2	29·16	13·08	27·9	22·72	5·9	22	·777	Full reports.
227	Melksham, Wilts	1870-5	5	4·27	47·	11·	33·	12·5	1·71	20	·53	
228	Warwick Cottage	1874-5	5	3·37	27·	8·	45·	88·88	1·06	..	..	Opened 1873.
229	Yoxall, Stafford	1870-5	5	1·5	10·83	7·22	39·26	0·	11·11	..	..	
230	Yate, Gloucester	1873-5	5	1·23	4·	3·25	112·5	0·	0·	..	..	Patients are paid for.
231*	Amlwch, Anglesey	1870-4	4	1·	10·3	10·3	34·3	100·	9·7	..	..	Reports complete.
232*	Bangor, Down, Ireland	1870-5	4	2·1	29·	13·8	26·3	55·23	8·	..	..	
233	Hillingdon, Middlesex	1870-5	4	·7	12·5	18·	19·37	42·85	5·9	19	3·105	Accident Hospital.
234	Scarborough, Accident	1870-5	3	·86	11·16	13·	28·09	76·72	6·9	21	3·289	
235*	Market Rasen, Lincoln	1870-5	2	3·5	72·5	20·7	29·7	142·85	8·55	..	..	
236	Leek, Staffordshire	1870-5	5	16·4	224·	13·65	23·88	122·92	..	..	..	
237	Kilmarnock Inf.	1870-5	..	..								Without fever cases, for which *see* Fever Hosp. (Too late for proper insertion.)

IRISH COUNTY INFIRMARIES, 1870—75.

* Hospitals from which Reports have been received.

No.	Name of Hospital	Years for which return is made.	Full No. of Beds.	Average Beds Occupied.	Average No. of In-patients.	Average No. of Patients to each Bed.	Mean Resi-dence.	Mortality, per cent. of Beds.	Mortality, per cent. of Patients.	District Mortality per 1000.	District Ratio of Hosp. to District Mortality.	Remarks.
1	Londonderry	...	128	...	...	...	...	...	...	...	...	Information promised.
2	Limerick	...	100	...	...	...	...	...	...	...	...	Information could not be supplied.
3	Maryborough	...	100	...	...	...	...	...	...	...	...	No reply.
4	Roscommon	1873–5	85	40·8	632·6	15·5	23·5	42·4	2·74	...	...	
5	Tyrone	...	85	...	...	...	...	...	...	...	...	No reply.
6	Galway	...	80	...	...	...	...	...	...	...	...	No reply.
7	Downpatrick	1870–5	80	59·	729·3	12·35	29·5	25·69	2·08	...	...	
8*	Armagh	...	72	...	...	...	...	...	...	...	...	No details in reports.
9*	Tipperary	...	72	...	...	...	...	...	...	...	...	No reply.
10	Wexford	...	72	...	...	...	...	...	...	...	...	No reply.
11	Cavan	...	70	...	...	...	...	...	...	...	...	No reply.
12	Donegal	1870–5	70	35·96	448·16	12·19	29·08	19·	1·52	...	...	
13	Kilkenny	...	70	...	...	...	...	...	...	...	...	No reply.
14	Clare	...	60	...	...	...	...	...	...	...	...	No reply.
15	Mayo	...	60	...	...	...	...	...	...	...	...	No reply.
16	Monaghan	...	60	...	...	...	...	...	...	...	...	No reply.
17	Sligo	...	60	...	...	...	...	...	...	...	...	No reply.
18	Fermanagh	1870–5	52	38·4	462·6	12·04	25·11	48·90	4·06	...	...	
19	Kildare	1870–5	52	25·8	291·8	11·3	32·3	20·54	1·81	...	...	
20	Leitrim	...	52	...	...	...	...	...	...	...	...	No fever cases.
21	King's County	1870–5	50	18·10	262·3	14·44	23·5	40·2	2·75	...	...	
22	Carlow	1870–5	42	...	262·4	...	...	...	4·57	...	...	No fever cases.
23*	Louth	...	40	...	...	...	...	...	...	...	...	Reports very defective. "Books not kept regularly."
24	Kerry	...	40	...	...	...	...	...	...	...	...	No reply.
25	Meath	1870–5	40	19·8	269·	13·56	26·9	36·16	2·66	...	...	
26	Westmeath	...	40	...	...	...	...	...	...	...	...	No reply.
27	Longford	...	39	...	...	...	...	...	...	...	...	No reply.
28	Wicklow	1870–5	36	9·	213·6	24·2	15·08	33·15	2·34	...	...	
	Total	...	1907	246·92	3571·76	14·44	25·27	33·25	...	...	...	
	Average ...	...	68·1	30·86	396·86	14·44	25·27	33·25	2·72	...	...	

FEVER HOSPITALS.

* Hospitals from which Reports have been received.

No.	Name of Hospital.	Years for which return is made.	Full No. of Beds.	Average Beds Occupied.	Average No. of In-patients.	Average No. of Patients to each Bed.	Mean Residence.	Mortality, per cent. of		District Mortality per 1000.	Ratio of Hosp. to District Mortality.	Remarks.
								Beds.	Patients.			
*	London Fever	1870-5	260	81·2	926·	11·4	32·	172·61	15·15	...	...	Metrep. Asyl. Dist. Board [Reports.
	Homerton	1872-5	105	...	85·25	...	...	...	18·	...	...	Do. Do. Do.
	Stockwell	1872-5	102	...	499·8	...	...	...	17·8	...	...	
	Fever, Cork	1870-5	140	81·16	704·5	8·68	31·	89·08	10·12	...	...	
	Cork Street, Dublin	1870-5	120	58·6	1041·	17·76	20·75	181·16	10·1	...	...	
	Hardwicke, Dublin	1870-5	120	51·63	750·	14·7	19·44	218·86	15·08	...	...	
	Leeds	...	80	...	...	...	...	...	...	...	...	No reply.
	Newcastle-upon-Tyne	1870-5	61	12·4	227·4	18·25	20·	26·74	6·9	...	...	
	Bradford	1872-5	60	19·	245·	12·87	28·34	205·26	15·87	...	...	No reply.
	Downpatrick	...	40	...	...	...	...	...	...	...	...	
	Halifax	1873-5	40	4·4	60·3	13·72	26·6	127·27	16·58	...	...	
	Queenstown, Cork	...	40	...	...	...	...	...	...	...	...	No reply.
	Newry, Armagh	...	40	...	...	...	...	...	...	...	...	No reply.
	Londonderry	...	42	...	...	...	...	...	...	...	...	No reply.
	Ayr	1870-5	32	3·74	51·4	13·72	26·6	22·55	11·68	...	...	
	Monaghan	1870-5	30	1·4	21·3	15·2	24·	130·71	9·	...	...	
	Carlisle	1870-5	26	...	79·5	...	...	...	13·85	...	...	
	Stafford Inf.	1870-5	...	...	165·	...	...	...	13·94	...	...	
	Manchester Inf.	1870-5	...	22·9	364·6	15·92	22·9	203·5	12·77	...	...	
	Arklow, Wicklow	...	16	...	...	...	...	...	...	...	...	No reply.
	Clones, Monaghan	1870-5	16	1·37	173·	12·6	29·	94·89	7·51	...	...	
	Altrincham, Cheshire	...	14	...	...	...	...	...	...	...	...	No reply.
*	Bedford	1870-5	50	...	525·	14·77	24·71	92·8	10·16	...	...	
*	Paisley	1870-5	...	20·7	305·5	...	...	...	7·3	...	...	

	Years									Corporation Hospital.
Homerton and Stockwell:										
Scarlet fever	1872–5			12·5				549·		...	...
Enteric fever				18·2				349·		...	...
Typhus				21·5				283·7		...	...
Average ...				17·6						...	...
Homerton Scarlet Fever	1871–5		45·8	7·	160·48	23·16	15·7	4275		...	...
Zymotics in Glasgow Inf.	1870–5		31·	10·	155·	23·62	15·45	720·3		...	...
,, ,, Dundee ,,	1870–5			10·				480·		...	...
Typhus ,, Dundee ,,	1870–5			12·8				216·		...	...
Typhus in Greenock Inf.	1870–5		8·18	13·28		21·3	17·13	140·2		...	...
Smallpox ,, ,,	1870–3		7·08	25·7		20·25	18·	127·5		...	...
London Smallpox	1871–5	108		18·5	227·38			635·8		...	...
Homerton and Stockwell Smallpox	1871			18·48	455·5			13087·		...	...
Birmingham Smallpox	1873–5	150	904·5	15·75						...	...
Sunderland Smallpox	1871–2			22·5				142·		...	...
Bedford Smallpox	1871			20·		25·	14·6	25·		...	...
Smallpox in Paisley Inf.	1874		20·7	14·86	82·8	18·	20·5	303·		...	...
Newcastle-on-Tyne Smallpox	1871–2		43·	19·35	397·			882·		...	...
Average ...				14·38		24·26				...	...
Smallpox...				19·91						...	...
Typhus ...				15·86						...	...

CHILDREN'S HOSPITALS.

* Hospitals from which Reports have been received.

No.	Name of Hospital.	Years for which return is made.	Full No. of Beds.	Average Beds Occupied.	Average No. of In-patients.	Average No. of In-Patients to each Bed.	Mean Residence.	Mortality, per cent. of		District Mortality per 1000.	Ratio of Hosp. to District Mortality.	Remarks.
								Beds.	Patients.			
*	Ormond Street, Lond.	1870–5	127	…	590·	…	…	…	10·	…	…	Reports defective.
	Evelina, London	…	100	…	…	…	…	…	…	…	…	No reply.
	Manchester, Pendlebury	…	84	…	…	…	…	…	…	…	…	No reply.
	Liverpool, Myrtle St.	…	80	…	…	…	…	…	…	…	…	No reply.
	Edinburgh	…	72	…	…	…	…	…	…	…	…	No reply.
	Birmingham	1870–5	50	46·38	774·5	16·68	21·88	102·1	7·14	…	…	Admirable reports.
	Victoria, London	1870–5	54	33·	315·	9·53	38·3	68·18	7·14	…	…	
	Bristol	…	50	…	…	…	…	…	…	…	…	Statistics too imperfect to be used.
*	Manchester Clinical	1870–5	46	…	…	…	…	…	…	…	…	No reply.
	Brighton	…	40	…	99·3	…	…	…	4·03	…	…	Reports defective.
	Gloucester	…	24	…	…	…	…	…	…	…	…	No reply.
	London Nth.-Eastern	…	24	…	…	…	…	…	…	…	…	No reply.
	Newcastle-on-Tyne	…	24	…	…	…	…	…	…	…	…	No reply.
	St. Joseph's, Dublin	…	21	…	…	…	…	…	…	…	…	No reply.
	Belfast	1873–5	20	9·3	24·26	26·	14·	60·21	2·3	…	…	
	Jenny Lind, Norwich	…	20	…	…	…	…	…	…	…	…	No reply.
	Belgrave, London	…	19	…	…	…	…	…	…	…	…	No reply.
	Wirrel, Birkenhead	…	16	…	…	…	…	…	…	…	…	No reply.
	Sunderland	…	…	…	…	…	…	…	…	…	…	
	Total … …	…	881	…	…	…	…	…	…	…	…	
	Average … …	…	49	…	…	…	24·72	…	6·12	…	…	

LYING-IN HOSPITALS.

* Hospitals from which Reports have been received.

No.	Name of Hospital.	Years for which return is made.	Full No. of Beds.	Average Beds Occupied.	Average No. of In-patients.	Average No. of Patients to each Bed.	Mean Residence.	Mortality, per cent. of Beds.	Mortality, per cent. of Patients.	District Mortality per 1000.	Ratio of Hosp. to District Mortality.	Remarks.
*	Rotunda, Dublin	1870-5	130	30·68	1237·	42·	8·2	44·32	1·057	...	...	Government Reports.
	Queen Charlotte's	...	50	...	...	...	...	...	...	...	...	No reply.
*	Coombe, Dublin	1870-5	40	10·62	396·	37·3	9·54	40·48	1·085	...	...	Government Reports.
	York Road, London	1870-5	30	14·8	300·8	20·3	18·	14·6	·717	...	...	
	Endell Street, London	1870-5	25	11·	174·16	15·83	23·	28·72	1·814	...	...	Mort. in 1786 was 1·285.
	Glasgow Maternity	1870-5	24	10·	315·6	31·56	11·5	51·6	1·635	...	...	
	Belfast	1870-5	15	...	185·3	...	...	...	1·07	...	...	
	Cork	1870-5	15	10·13	370·	36·5	10·	...	·225	...	...	
	Limerick	...	12	...	...	...	...	...	...	...	...	No reply.
	Edinburgh	...	10	...	...	...	...	...	...	...	...	No register kept.
	Waterford	1870-5	8	...	120·3	...	...	...	·25	...	...	
*	City of London	1870-5	30	27·3	406·	10·09	19·16	23·8	1·6	...	...	
	Marburgh, Hesse	...										
	Cassel	1868-75	30	...	129·	...	...	...	1·162	...	:	
	Average ...	...	...	...	...	...	11·34	...	1·061	...	...	

WOMEN'S HOSPITALS.

* Hospitals from which Reports have been received.

No.	Name of Hospital.	Years for which return is made.	Full No. of Beds.	Average Beds Occupied.	Average No. of In-patients.	Average No. of Patients to each Bed.	Mean Residence.	Mortality, per cent. of Beds.	Mortality, per cent. of Patients.	District Mortality per 1000.	Ratio of Hosp. to District Mortality.	Remarks.
*	Soho Square, London	1870-5	60	37·	337·	9·1	40·	51·35	5·7	...	...	
*	Samaritan, London	1870-5	50	...	252·	...	...	...	5·28	...	...	
	Waterloo Road, Lond.	...	50	...	...	...	...	...	·67	...	...	
*	Leeds	1870-5	45	19·	294·6	15·5	23·5	10·52	1·8	...	...	
	Marylebone Rd., Lond.	1872-5	26	10·4	137·25	13·4	28·	24·	...	...	...	
	Samaritan, Belfast	...	25	...	...	...	...	...	...	...	...	
	Cork	1870-5	20	29·16	312·5	10·7	37·3	2·05	·16	...	—	
	Sheffield	1873-5	12	7·3	77·	10·55	34·6	41·1	3·9	...	...	Opened 1875.
	Vincent Sq., London	1873-5	10	4·3	30·	7·	53·6	30·	4·35	...	...	Govt. Lock Hosp. under Contagious Diseases Act.
*	Birmingham	1871-5	8	2·44	615	2·52	15·	204·91	8·1	...	...	
	Chelsea	...	8	...	...	...	...	...	...	...	...	No reply.
	Newcastle-on-Tyne	...	4	...	...	...	...	...	...	...	...	No reply.
	Average	...	...	...	...	...	33·	...	37·4	...	...	
	Sanatorium for Consumptives, Bournemouth	1870-5	46	46·5	230·2	4·95	73·73	6·02	1·21	...	...	

Summary of all the Returns obtained from 1870—75.

N.B.—The divisors for these columns are not constant.

Hospitals		Full No. of Beds.	Average Beds Occupied.	Average No of in-patients.	Average No. of Patients to each bed.	Mean Residence.	Mortality per cent of — Beds.	Mortality per cent of — Patients.	District Mortality per 1000.	Ratio of Hosp. to District Mortality.
Hospitals having from 2 to 20 beds	Total	1367	272·91	3860·16						
	Average	9·49	4·01	48·25	15·2	...	60·41	5·17	20·2	2·56
Hospitals having from 21 to 99 beds	Total	4333	1132·49	15710·46						
	Average	48·14	24·6	270·87	10·76	24	69·77	6·60	21·72	3·041
Hospitals having from 100 to 199 beds	Total	7204	3894·8	47091·06						
	Average	131	97·37	961·04	11·00	33·92	81·22	7·00	23·52	2·976
Hospitals having 200 beds and over	Total	9034	5552·16	64352·84						
	Average	347·46	231·34	2681·36	12·04	33·16	94·53	7·98	28·	2·85
All:	Total	21938	10852·36	131014·52						
	Average	73·12	52·46	632·92	12·52	30·31	70·97	6·24	22·10	2·816
20 English County, Nos. 17, 21, 31, 33, 46, 48, 49, 52, 55, 56, 59, 60, 64, 65, 66, 70, 72, 74, 76	Total	2611	1717·64	17311·86						
	Average	130·55	95·42	865·59	9·07	29·15	53·47	5·21	22·3	2·239
6 Irish "Large Town," Nos. 15, 18, 26, 35, 54, 67	Total	1052	664·52	10589·4						
	Average	175·3	132·9	1764·9	13·28	40·24	90·55	6·40	...	...
14 English "Large Town," Nos. 8, 11, 13, 14, 19, 22, 23, 36, 37, 40, 42, 43, 44, 45	Total	3050	1793·55	23056·2						
	Average	203·3	149·38	1819·7	12·37	27·48	96·94	7·85	29·93	2·624
7 Scotch Infirmaries, Nos. 4, 6, 10, 12, 24, 30, 47	Total	2243	1445	18197						
	Average	320·4	206·4	1697·32	11·74	31·09	115·60	9·53	...	...
8 Accident, Nos. 105, 117, 137, 158, 162, 173, 220, 234	Total	160	43·73	6307·25						
	Average	20	6·29	242·47	11·88	33·13	138·64	9·68	22·4	4·718
14 London, Nos. 1, 2, 3, 5, 7, 9, 25, 32, 34, 38, 41, 68, 71, 132	Total	4200	3018·44	36665·18						
	Average	323	232·18	2576·08	11·09	32·91	118·47	9·88	23·93	4·212

Summary of those Hospitals (179) from which complete information was obtained, arranged in groups according to the number of Beds in average occupation.

No.		Average Beds Occupied.	Average No. of Patients to each Bed.	Mean Residence.	Mortality, per cent, of		District Mortality per 1000.	Remarks.
					Beds.	Patients.		
I.	54 Hosps. under 5 beds	2·82	12·	34·5	62·92	5·03	20·14	
II.	12 ,, from 5 to 9 beds	6·62	11·14	32·03	58·97	5·02	21·25	In this " 3rd " group there are 6 Hosps. with an average mortality of 11·57; and if these were eliminated the average mortality of the 20 others would be 5·636, and the bed rate 61·26.
III.	26 ,, 10 ,, 19 ,,	13·44	11·3	32·23	79·77	7·00	21·	
IV.	8 ,, 20 ,, 29 ,,	25·	9·4	39·65	46·16	5·63	21·	
V.	6 ,, 30 ,, 39 ,,	33·96	9·88	35·12	58·09	5·94	22·3	
VI.	8 ,, 40 ,, 49 ,,	43·77	9·6	40·17	57·06	5·97	22·3	
VII.	11 ,, 50 ,, 74 ,,	61·82	10·35	36·33	61·69	5·77	21·8	
VIII.	15 ,, 75 ,, 99 ,,	·84·05	11·08	34·26	77·54	6·66	24·13	
IX.	10 ,, 100 ,, 124 ,,	113·35	11·85	32·48	79·18	6·7	24·	
X.	17 ,, 125 ,, 199 ,,	143·74	11·95	31·46	93·34	7·92	24·92	
XI.	5 ,, 200 ,, 299 ,,	211·91	12·39	30·18	111·85	9·15	30·6	If the mortality were the same in XII as it is in XI. (that is, ·443 less), 140 lives a year would be saved.
XII.	7 over 300 beds	417·3	10·74	33·38	101·53	9·6	25·	

If it be true, as all statisticians hold it to be, that the employment of large masses of figures enables us to get rid of minor sources of error, I think I may fairly say that the returns now before us afford as reasonable a basis for estimating what really is the mortality of our general hospitals as can be obtained. Approximately more exact results would have been yielded if all the hospitals could have been included; and I can but at the least hope that the publication of my tables will induce hospitals generally to keep their records more exactly, and to publish in their reports such details as will enable their comparative and absolute utility to be rightly estimated.

From the general summary of my results, it appears that the average of the full number of beds returned by all the hospitals is 73·12; and that the average of the number of beds occupied is 52·46. There is, therefore, a constant margin of about 50 per cent. of beds over and above those constantly occupied, or at least room for such a number of beds.

This is probably largely in excess of what is actually the fact; but it is perfectly evident from the weekly fluctuations of our hospital population, that a considerable margin of beds must be kept in readiness for emergencies over and above what corresponds to the average population.

There rises here a very important question as to whether the extent of this margin may not very materially affect the death rate of the hospitals; for in hospitals where the margin is very narrow, as University College and the Sheffield Infirmary, the mortality is very high; whilst in others, where the margin is large, as St. Bartholomew's and the Leeds Infirmary, the mortality is very much lower.

Yet the two pairs of hospitals quoted are in other respects quite fit for comparison. We have only to look at the reports and see that the work at St. Bartholomew's is quite as active as that at University College; and there is no explanation to be found in the comparative death rates of Leeds and Sheffield to tell us why the hospital mortality of the one should be nearly double that of the other. Of course, in any hospital where there is a large margin of unoccupied beds, it follows that probably a larger amount of floor and cubic space is allowed for each of the average inhabitants. But other disturbing elements come in; for at St. Thomas's Hospital, where there is a very large margin of unoccupied beds, in a new hospital, the mortality is higher than either at Sheffield or University College. It surely rests with the authorities at St. Thomas's to show that this is not due to the closing of some wards and the overcrowding of others, or to any other cause which is removable.

The district mortality of St. Bartholomew's is higher than that of St. Thomas's, so that *à priori* there seems no reason why there should be such a great difference in the death rates of the two hospitals. If the death rate of St. Thomas's was as low as that of St. Bartholomew's, 220 valuable lives would be saved every year.

The benefits of a large margin of hospital space obtained by reducing the constant population is seen by contrasting the results of St. Bartholomew's Hospital for the years from 1861 to 1864, published by the committee of the Statistical Society, with those in my own tables. During the first period, the hospital had an average population of 547 patients, with a mean residence of 37 days, there being only 9·86 inhabitants for each bed per annum. The bed rate was 108·592, and the patient rate 10·822 per cent., the marginal bed accommodation being only 19 per cent. of the whole. From 1870 to 1875 the average population was brought down to 301·49, having a marginal accommodation of 57·5 per cent. The patient rate has fallen to 5·128, and the bed rate to 56·91 ; that is, that while the latter used to be in the ratio to the former of almost exactly 10 to 1, it is now more than 11 to 1, showing that the beds are more useful by 10 per cent. than they formerly were— a conclusion which is borne out by the fact that the mean residence has fallen 4·15 days, and 11·11 patients are now treated in every bed per annum.

The real improvement in the work of the hospital must be much greater than these figures can show ; for as the total number of in-patients has been reduced nearly 40 per cent., it is certain that the most important cases will be selected for admission. This is an example worthy of imitation, and surely these facts alone are enough to cause the whole question of hospital mortality to be subjected to the scrutiny of a scientific commission.

The next conclusion which is pointed to by my returns is that for every bed constantly occupied there are 12·52 patients during the year, which gives a mean general residence of 29·15 days for every patient. From this I think it may be very fairly concluded that these two figures are hospital constants of great value, from which may be approximately determined the relative activity of the work done in hospitals. There can be no doubt that prolonged residence in a constantly-occupied building must have evil results. This is so constantly seen amongst human beings in health, that it requires no argument to support it as a proposition concerning disease. But it will afterwards be seen that the figures from certain classes of hospitals point out this incontestably.

In her answers to questions asked by the Royal Commission, given in her " Notes on Hospitals," Miss Nightingale states that in the Scutari Hospital the average residence was 39 days when the mortality was 31·5 per cent., and only 24 days when the mortality fell to 2·2 per cent. This is very strong evidence, and even allowing a margin for possible error, it is almost conclusive that conditions which lead to a high death rate almost certainly diminish the residence, even when the chief causes of mortality are such diseases as run short courses.

Thus zymotic hospitals have the highest death rate and the shortest mean residence. Such figures as those given by Miss Nightingale were for cases amongst which wounds of course greatly predominated, and in these cases the unsanitary conditions would of course at once raise the death rate and prolong the residence of the survivors.

The three columns of " Average patients to each bed," " Mean residence," and " Mortality per cent. of beds," taken together, will be found to determine very fairly the amount of usefulness of any hospital ; and if taken along with the last three columns, a very distinct indication is given of undue mortality. The column of bed mortality shows that the average number of deaths per hundred beds is 70·974 ; or that every bed in average occupation in every general hospital will have ·709 of a death occur in it during a year.

When the bed rate falls greatly below this there is reason to suspect a deficiency in the usefulness of the hospital ; and when it is greatly in excess of it, as at the Manchester Infirmary and University College Hospital, where it is more than double, there is more than reason to suspect that the hospital has some intrinsic cause of unhealthiness. I think this must be held to be especially the case where, as at University College, the mean residence is not much below the average. Contrasting the facts of this hospital with an exceptionally bad zymotic hospital, such as the Smallpox Hospital in Greenock and a Children's Hospital as that of Birmingham, this conclusion is made almost certain.

At Greenock the residence falls to 18 days, and the bed rate rises to about 455·5; at the Birmingham Children's Hospital the residence is 21·88 days, and the bed rate only 102.

The exact and immediate causes of this excessive bed rate at University College, coincident with an almost average residence, cannot, of course, be determined by a mere inspection of figures, still less could any suggestion for remedies be obtained from them. But they point out a state of matters meriting a most searching investigation.

My last two columns were drawn up in the hope that they would point to some much more definite conclusions than they seem to do. Still it may yet not be without importance to know that the average death rate of general hospitals is 2·816 times the death rate of the districts in which they are situated.

I have further summarised my returns in two tables, in the first of which I have placed the hospitals in four groups, separated by very artificial and not very satisfactory lines of demarcation, but adopted because Simpson has taken a somewhat similar method of comparing hospitals by their amputation returns. I may here express my conviction that Simpson's conclusions are far from being as yet substantiated, but they are so probable that they may be at least provisionally accepted as indicating a large measure of truth. His method was inexact and open to important objections, but it will be seen that my figures lend his conclusions very strong support.

I have in this table taken out the returns of all the hospitals having twenty beds, and under. I find the average full number of beds is 9·49, and that the number constantly occupied is less than half. This is to be expected in small hospitals, as they are much more liable to have their accommodation tried by emergencies than are large hospitals. Thus no accident or barometric variation is likely to involve such a number of victims as greatly to try the marginal accommodation of a hospital of more than 200 beds. But an accident affecting five people would greatly strain the powers of an average hospital under 21 beds. In this class of hospital the number of patients per bed is increased 2·7 over the general average, and the mean residence falls 5·15 days. We have no clear evidence of the cause of this. Those who advocate small hospitals will say that the patients recover better and leave sooner. Those who apologise for large hospitals will say that the cases admitted are more trifling than those admitted to town hospitals. Mere expressions of opinion are of but little use in an inquiry like this, but I cannot help saying that the reports of very many of these small hospitals give full details of every case treated, and I have failed to be convinced, after carefully reading a large number, that the cases are less severe than those admitted into town hospitals. That severe cases recover better there than in the town hospital is a matter quite beyond dispute, but the causes of the better results are beyond the reach of figures.

The bed and patient death rates in these hospitals are both below the average; and the ratio of the district to the hospital mortality is only as 1 is to 2·56. I think that this tends to show that the gain in salubrity is due to intrinsic causes.

In hospitals having from 21 to 99 beds the average full number is 48·14, and the constant population fills just one half of them. The number of patients to each bed falls 1·76 below the general average, and the residence rises 4·77 days. At the same time the bed rate is slightly under the general average, and the patient death rate slightly over—results which are unquestionably due to defective management in this class of hospitals, which includes a large number of county and small country town hospitals. A glance only has to be given to the returns from the Essex and Colchester, the Taunton and the Dorchester hospitals, for evidence of this. The abuse of these hospitals is unquestionably due to the almost uniform prevalence amongst them of the ticket system, a system which is as demoralising to those supposed to receive benefit as it is discreditable to those who are supposed to be dispensing charity. In these hospitals, though the bed rate is lower and the patient rate only slightly higher than the general average, the fact that the district mortality, 1·62 less than the average, stands to the hospital mortality as 1 is to 3·0·11, seems to indicate that the patient mortality is to be really considered excessive. Here is another point which demands a rigid scrutiny ; for it must also be suspected that in many of these hospitals the public money must be wasted to a very unnecessary extent. It is worthy of notice that at the Bridgewater Infirmary all in-patients are made to pay a small weekly sum towards their maintenance, and to this the authorities of the hospital attribute the greatly diminished period of residence, 19 days. The same plan is in

use at the Erith Cottage Hospital—an institution which, it
seems to me, may serve as a pattern for most institutions of
the kind.

The next class of hospitals, those having from 100 to 199
beds, includes some of the most important in the country,
and it must indeed be said for them that they are far from
being clear of the defects of the class immediately preceding them. The mean residence is high, and it is only
when those having over 200 beds are counted with them
that they seem to present comparative activity. The
diminished mean residence is then, however, only the result
of increased mortality, for while it sinks to only 1·165 days
above the general average, there is an excess of 23·558 in
the bed rate, and 1·732 in the patient death rate. The
district mortality rate in this class rises to 1·7 in 1000 above
the general average, and the ratio between it and the hospital rate remaining very much the same, it would seem as if
the general mortality exercises a constant influence on the
hospital rates, as might be expected, and as was formerly
suggested by Dr. Guy.

In the second table of summary I have arranged those
hospitals, 179 in number, from which I have obtained complete information, into twelve groups; and by using the
average number of beds of each group as the abscissa, and
various factors as ordinates, I have constructed a series of
curves which are very interesting, and which show at a
glance a number of facts referred to in the letterpress. It
is especially evident that the hospitals in the 4th, 5th, 6th,
and 7th groups are not managed with sufficient stringency
in the matter of residence; that at the 10th group the increasing mortality brings down the average residence; and
further, that from this point the still increasing mortality is

Nº of Beds. — District Mort. — Hospi Mort. — Patients per Bed — Residence in days

40 39 38 37 36 35 34 33 32 31 30 16 15 14 13 12 11 10 9 8 7 6 5 4 3 2 1

2.80 6.62 13.44 25 43.77 61.82 84 113.35 144 211.9 417.3

I II III IV V VI VII VIII IX X XI XII

combined with a prolonged residence and diminished number of patients per bed,—these three curves proving an increase in the unsanitary condition of hospitals as they increase in size.

When we come to consider the London hospitals as a class by themselves, we find some figures which are certainly surprising, and may form a fruitful subject for speculation. First of all we find that 71.8 per cent. of the full number of beds are in constant occupation, leaving a margin of accommodation of only 28·2, a probable source of general unhealthiness. The number of patients to each bed is only 11·09, a sad falling off in usefulness when compared with any of the figures except those of the third class; and even then there is only ·085 of a patient in favour of the London hospitals. The residence is very high—nearly 33 days, or 3·76 above the average. The bed rate is terribly high, and the patient death rate is 3·63 above the total average. Further, the ratio between the district mortality and the hospital rate is as 1 is to 4·212. Into the causes of these facts I cannot, as I have said already, enter here. But I must say it will require a very weighty amount of evidence to convince me that they are inevitable and irremovable.

We have been told in a Parliamentary report on the subject of hospitals, that the method of admission is an important factor in hospital mortality, and that the ticket system keeps down the death rate. The Royal Free Hospital is absolutely devoid of the ticket system, and the cases are admitted to its wards by reason of their urgency, yet the patient death rate is only 6·968 or 2·913 below the average of the London hospitals. At the University College Hospital the ticket system, according to the reports, is still in existence, yet the death rate is 11·91. We have further been told that the reputation of the staffs of particular hospitals, by bringing desperate cases, raises the death rate ; and that on this account, as well as for other reasons, a high death rate is the only test of a hospital's usefulness. This is nonsense, for no hospital staff in London or elsewhere has, for the last thirty years, had such a reputation as that of St. Bartholomew's, and the reports issued by that staff show work done which challenges comparison with that of any hospital in the world, and yet the death rate of beds and patients at this hospital is the lowest in London.   Besides, it cannot be said that the reputation of the staff diminished so much in ten years, from 1864 to 1874, as to lower the mortality at St. Bartholomew's 50 per cent.   The statement, also, that the connexion of a medical school with a hospital must necessarily increase its death rate is also met by the returns of St. Bartholomew's, St. George's, and Guy's.

The largest mean residence in the London hospitals is found at Middlesex, a fact probably due to the three cancer wards which, I presume, are pretty constantly occupied by patients who have a very prolonged average residence. The system, however, of having such cases in a general hospital is one of very doubtful propriety, and no details are given in the reports of this hospital from which conclusions can be drawn.

The figures in my tables do not show that the mortalities of the London districts exercise a constant influence on the hospital mortality, as has been asserted, unless perhaps we conclude that the same circumstances which conduce to the low mortality of the districts in which St. George's

Hospital stands, also mitigate the death rate within its walls.

Mere size does not seem to have a constant influence in raising the death rate of a hospital, otherwise Guy's would have a greater mortality than University College, whilst really it is 2·41 per cent. less. It is of course more likely that a large hospital will prove more unhealthy than a small one, because the chances of having known causes of high death rate in existence are greater when a large number of people are gathered together in a large building, than where a small number are collected in a small building. But it would be quite as easy to smother people in the cabin of a canal boat, as it was in the black hole of Calcutta. The numbers killed would not be so large, but the death rate might be made far higher. That a small hospital can be made quite as unhealthy as one of the largest size has been often and abundantly proved.* It has been also shown in the cases of all kinds of human habitations, hospitals as well as others, that causes of unusually high death rates are almost always recognisable and removable. I am of opinion, therefore, that in the case of every hospital when the mortality is found to be unusually high, that it is incumbent on the managing body to show what the causes are which are not removable, and to remove at once those which are. Further, I think that the medical officer of health for

---

* This is well shown in the graphic sheet. In the third group there are 26 hospitals, 20 of which have an average mortality of 5·636, or just what the mortality of the whole ought to be. That, however, is raised to 7 per cent. by the other six hospitals—Lancaster, Whitehaven, Chesterfield, Burton-on-Trent, Oldham, and Doncaster. It is quite impossible to believe that these six hospitals must necessarily have an average mortality of 11·57 per cent. The Whitehaven Infirmary admits fever, and has a mortality of 14·36 per cent., a state of matters demanding instant remedy. The Oldham Infirmary also requires attention, for its mortality is quite as high.

every district should be made to act as a statistical auditor
for the hospitals in his district. It will be borne in mind
that, a few years ago, the lying-in department of King's
College Hospital had to be closed on account of the
terrible mortality among the parturient women, due to
causes intrinsic to the hospital. Are its managers in a
position to show that its high mortality—12·05 per cent.,
the highest in London except St. Thomas's, and quite as
high as many fever hospitals—is not in any way due to the
same causes as those which lead to the outbreaks of puerperal fever ?

It has been repeatedly stated, merely of course as a matter
of opinion, for no figures have been published to show the
facts, that the excessive hospital mortality in London is due
to street accidents and patients coming from the country.
But if the reports of London hospitals be compared with
such a hospital as the Leeds Infirmary, it will be seen that
accidents in London are mere trifles compared to the accidents in a purely manufacturing town, and that Leeds is
comparatively quite as important a centre to which patients
gravitate from the surrounding districts as is any London
hospital. In fact, as far as the facts can be determined,
the cases from a distance which seek relief in London are
chronic cases which generally have filtered out of local hospitals without receiving much, or at least permanent, benefit.
These are not cases in which the mortality can be very high,
though I cannot state it in figures. Acute cases cannot go to
London, and it is these which run up the death rate. We
find then that the entire mortality of the Leeds Infirmary
is 3·101 less than the average of all the London hospitals,
and not much more than half those of St. Thomas's,

Middlesex, or King's College. In the Leeds Infirmary we find that there is a margin of 30 per cent. of unused beds. By the courtesy of the secretary of this admirable institution I have been put in possession of the measurements of its wards, and I find that to each constant patient a square area of 106 feet, and a cubic space of nearly 3000, is allowed. This is a new hospital, and as I fortunately am in possession of the returns for the last ten years of the old Leeds Infirmary, an interesting comparison may be made. I have not any measurements of the old hospital, but I knew it well, and remember very vividly the low roofs, the proximity of the beds, the defective ventilation, and the close smell of the wards. Ventilators had been provided, but they were found to be used chiefly by birds for building purposes. There was a nominal margin of 25 per cent. of accommodation not constantly used ; but if the whole number of beds had ever been in use, some of them must have been as hammocks slung from the roofs. The mean residence was 32·56 days, whilst in the new infirmary it is 25·625. In the old building the bed rate was 91·733, in the new it is 100, whilst the patient death rate has fallen in the new hospital 1·499 per cent. From this we may conclude that nearly 60 valuable lives are saved every year by the new Leeds Infirmary ; and I have reason to know that the authorities of the hospital are persuaded that the present death rate may be still further reduced.

Now, as the constituency of the Leeds Infirmary has been, during the last six years, drawn from the same population as it was for the ten years preceding. the kind of work done could have been in no way different. The staff has not been materially altered, the same amount of skill, and the

same appliances have been employed, so that the conclusion
is irresistible that the diminished mortality is due to the im-
proved sanitary conditions under which the patients are
treated, and chiefly to a greater allowance of bed and
breathing space.   It is clear, then, that it is the duty of
every hospital having an excessive death rate to see whether
that cannot be diminished by lessening the constant population, thereby increasing the space allowed to each of its inhabitants.   Howard spoke of the Leeds Infirmary as one of
the best hospitals in the kingdom at that time.   " Wards 15
feet 6 inches high, great attention to cleanliness, no fixed
testers, no bugs.   Many are here cured of compound frac-
tures, who would lose their limbs in the unventilated and
offensive wards of some other hospitals."   If he had lived
in our own day he would have spoken of the new hospital
quite as highly in comparison with some others now in
use.

In another group I have placed together fifteen of the
hospitals of the large towns of England.   The aggregate
district mortality of these towns is 6 in the 1000 higher than
the mortality of London, and yet the average mortality in
their hospitals is 2·05 per cent. less than the average London
hospital mortality ; and the mean residence is 3·41 days less
in the provinces than it is in London.   Of the whole 15
there are only two, the Manchester Royal and the Sheffield
General Infirmaries, where the London hospital mortality is
approached.   Fever cases are admitted at Manchester, a
practice which, for a town of that size, is wholly unnecessary,
quite indefensible, and one which ought at once to be discontinued.   But besides this, there must be at the Manchester
Infirmary some serious intrinsic causes of the high death

rate. The mean residence is 25·8 days, the bed rate is
152·46, and the ratio of the district mortality to that of the
hospital is as 1 is to 3·37. These facts are quite enough to
justify the course the governors have taken in asking for the
inspection of the hospital by an expert. I think there can
be little doubt that it would be better in every way to
remove the greater part of the hospital practice to the
outskirts of the town, in six hospitals of 100 beds each,
retaining in the centre only such accommodation as is
required for cases of emergency.*

I have not been able to obtain any information as to the
causes of the very high mortality at the Sheffield General
Infirmary. The very narrow margin of accommodation is
highly suggestive that the hospital is greatly overcrowded.
Its mean residence is much above the average, and its bed
rate is therefore comparatively low. There is another institution in Sheffield, called the " Public Hospital," but its
reports are so defective as to be quite discreditable to a
public institution, and its returns are so open to suspicion,
that the results given—8·14 per cent. of patients for two
years—cannot be taken as a basis of comparison with the
11 per cent. mortality of the General Infirmary.

But when two or more hospitals, existing in the same
town, give accurate returns, they may be fairly contrasted.
In Liverpool there are three large hospitals—the Infirmary,
and the Northern and Southern Hospitals.

The Infirmary is the largest of the three, and it has the
highest death rate and by far the longest mean residence,
yet both the Northern and Southern Hospitals seem to have
a much larger proportion of surgical cases, especially of

---

* Since this was written, Mr. Netten Radcliffe's report has abundantly proved
the correctness of my suppositions regarding the Manchester Infirmary.

accidents, than the Infirmary; so that we should expect
them to have a longer mean residence. In the Liverpool Infirmary a number of insane patients are treated—
a practice which seems to me open to very great objections, but these cases are not included in my returns.
The mean death rate of these three hospitals is 6·486, and
the general death rate of the town is 39 in 1000,
giving a ratio of the latter to the former as 1 is to 1·66.
The average death rate of London is 23·93, and the hospital rate is 9·88, giving a ratio of 1 to 4·212. This
remarkable difference between the conditions of the two
towns is of course in great part due to the high general
death rate of Liverpool, but the great difference between
the hospital death rates in favour of the latter town is an
equally important factor, and is strongly suggestive of the
necessity for a very searching inquiry into its causes.

A very striking improvement in the returns of the Southern
Hospital, due apparently to a change of the building, is quite
comparable with what I have shown as taking place in Leeds.
In 1870 the full number of beds was returned as 120, the
average number occupied being 89·3, the margin of accommodation being therefore 25 per cent. of the whole. In
1873 a new hospital was opened, the full number of beds of
which is 200. Of these 124·3 are occupied on the average,
leaving a margin of nearly 38 per cent. The number of
cases treated has greatly increased, the mean residence has
fallen nearly a day, and the patient death rate has diminished
1·76 per cent. There has been, as in the case of the Leeds
Infirmary, no change in the constituency; and it must be
shown, not merely stated, that the character of the cases has
become more trivial. I presume the new hospital was built
because the old one was found hurtful to the patients, and

it is a cause of congratulation to the managers to find that their efforts have succeeded apparently in saving 35 more lives every year.

In Birmingham there are two large hospitals, the Queen's and the General Hospital. The mortality of the former is ·4 less than that of the latter, the mean death rate of the two being 7·872. The general death rate of Birmingham is 12 in 1000 less than that of Liverpool, yet the hospital rate in the latter town is 1·386 per cent. less than that of Birmingham—a fact which I think should induce the managers of its two hospitals to consider very carefully whether they may not be able to reduce their mortality bills very materially. This could probably be done by a reduction of the constant population of the hospitals; for at the Queen's Hospital the marginal accommodation is only 14 per cent., and the mean residence is 29·5 days; whilst at the General Hospital the margin is 19 per cent., and the mean residence only 27·8 days. This considerable difference in the residence is not explained by the slightly higher mortality at the General Hospital; and as the admissions are chiefly privileged, the shorter residence is highly creditable to the management of that institution.

In Bristol the mean residence at the General Hospital is 1·25 days below the average; the patient death rate is 6·6, and the marginal accommodation is 28 per cent. of the whole. The condition of this hospital seems to be generally satisfactory, and it is a matter of regret that I am unable to contrast it with the Royal Infirmary in the same city.

At Newcastle Infirmary there is 32 per cent. of marginal accommodation, yet the residence is 4·93 days above the average, and the death rate is 8·17—a state of matters which is eminently unsatisfactory. The same may be said of Hull, a town with the comparatively low general death rate of 26 in 1000. The mean residence in the infirmary is 2·1 days above the average, and the patient death rate is 8·93.

In Bradford the residence is higher than in any other large town, being 39·5, or ten days above the average; and with quite an average death rate, a large margin of beds, and the very low bed rate of 72·61, it is evident that there is a want of vigilance in the executive of this hospital in seeing that patients are not kept in too long.

I have grouped together seven infirmaries in the large towns of Scotland; and as the reports of these institutions are generally very full, some being so complete that they may serve as models, they are easily compared.

They are nearly all absolutely free institutions—that is to say, subscribers do not have, or do not exercise, any privileges of presentation. Of these the highest mortality is found to exist in the Greenock Infirmary—an institution and a town both of which really ought to be the subject of a special inquiry. More than half the cases treated in this hospital are zymotics, and the mortality of all the cases is 12·24, whilst it actually rises to 12·56 if the fever cases are excluded. This terrible result may be due solely to the treatment of such a mass of zymotic disease in association with cases of other diseases, but there is strong ground for suspicion that the hospital is otherwise unhealthy. It cannot be doubted, I think, for a moment, that the fever hospital ought to be removed at once to the outside of the town, and the whole place which produces such a mass of zymotic diseases should be declared insanitary under the recent Act for the purpose of dealing with such areas.

If the statement of the Greenock zymotic cases, given

amongst the fever hospitals, be examined, it appears that there is in that town a hospital mortality which approaches more nearly to the dreadful experiences at Scutari than anything else that I have been able to collect. The explanation of this state of matters must be within reach, and the interests of humanity demand that it should be discovered and such steps be taken as will lead to a radical alteration of the hospital death rate. It is not easy to understand why, in a country like Scotland, where vaccination has been compulsory for very many years, it should be necessary for one case of small-pox in every four to die; whilst in a town like Birmingham, where, until very lately, vaccination has been but carelessly attended to, only one case in six and a half should prove fatal. The difficulty is quite as great in reconciling the enormous hospital zymotic death rate at Greenock of 16·7 per cent. with that seen in the neighbouring town of Paisley of 7·3. It is perfectly true that some of the excess may be due to importation amongst the floating population of Greenock, but similarly high mortalities are not found to occur in other seaports. If the results at Greenock be compared with those of Paisley, it will be seen that the death rate is not greatly modified, whether we include or exclude the zymotic cases— a fact which is to me very unexpected. The total death rate in the Paisley Infirmary is 4·26 less than at Greenock, and the zymotic cases at the latter hospital are 36 per cent. in excess of those at Paisley. The conclusion is inevitable that the admission of these cases has a great deal to do with the increase of the death rate of the non-zymotic patients; for there is found to be, in both infirmaries, almost no difference in the death rate, whether the zymotic cases be included or not. In the Dundee Infirmary a

large number of zymotic cases are treated, and their death
rate is 2·8 per cent. higher than that of the non-zymotic
cases.   The zymotic death rate is not quite so high as that
of Greenock, but it is much higher than that of Paisley.
All three towns require treatment as insanitary areas, and
the death rates of the hospitals of Greenock and Dundee
are specially worthy of-the attention of their managers.

Of course in these three hospitals the same proportions of
the various zymotic diseases do not exist, and to compare
the results of individual diseases in these institutions would
be impossible in the scope of this work.   But I have taken out
the typhus deaths, and I find that the mortality rate of
that one disease is higher at Greenock than at Dundee.
Bearing in mind that this is a disease which may be entirely
prevented, the mortality displayed is terrible.

The Scotch infirmary which has the most creditable results
is that at Aberdeen.   It may be said that this hospital is
hardly comparable with the infirmaries of Edinburgh and
Glasgow, but I cannot discover any reason in the conditions
of the population of Edinburgh, still less of the wider constituency from which the patients are in great part drawn,
which would account for a difference of 4·58 per cent.
between the hospital mortalities of Edinburgh and Aberdeen.   The former city has a great advantage in not being
a manufacturing centre, and it has an unrivalled situation.
The infirmary is well placed for ventilation, but it has
always borne an unenviable character for hospital diseases.
A special system for the treatment of surgical cases in-
volving operations has been introduced here during the last
six years, and it has been greatly praised by its advocates.
The hospital reports show that it has added many hundreds

of pounds a year to the drug bills. It does not seem to have had any appreciable influence on the statistics. If its success had been as great as it has been stated, it would have materially prolonged the mean residence in the hospital, because it is applied solely to the surgical cases, and it would also have materially decreased the mortality. Neither of these effects, however, is apparent. The mean residence in the Edinburgh Infirmary is 3·64 days longer than at Aberdeen, and 2·4 days in excess of that at Glasgow—conditions which are suggestive that patients recover more rapidly at Aberdeen, and that they are more likely to die at Glasgow. At the latter hospital the high mortality, 11·1 per cent., along with the mean residence of 30·6 days, should excite very especial attention. The reports of the Glasgow Infirmary are admirably full, whilst those of Edinburgh are so meagre that it is impossible to make any detailed comparison between the two institutions. It is, however, remarkable that at Edinburgh there is a margin of accommodation equal to 27·2 per cent., whilst in Glasgow the margin is only 18·6 per cent. The difference between the mortalities of the two is ·92 per cent., in favour of Edinburgh. It is possible that the difference in the marginal accommodation may help to account for this ; and I am quite sure that it is not to be explained by the mere statement that in the case of Glasgow we have to deal with a large manufacturing centre ; for if this were the important factor, it should not have a higher hospital mortality than Liverpool, Birmingham, or Bristol.

On the whole, the Scotch infirmaries contrast favourably with the London hospitals, and they do this in spite of the fact that they all receive a large, in some instances a very large, percentage of zymotic diseases. Typhus is admitted abundantly, whilst in the London hospitals it is at least very rare, if not altogether unknown in most of them. The mean residence in the Scotch infirmaries is nearly two days less than that in the London hospitals, and the patient death rate in the former is ·351 less than in the latter; so that we may conclude that in Scotland the infirmaries are both healthier and better managed than the hospitals in London. I think the former institutions would greatly improve their death rate if they would treat their zymotic cases at a distance from the general hospitals.

My next group includes six large general hospitals in Ireland—five in Dublin and one in Belfast. I must here say that in dealing with the hospitals in Ireland, great caution must be exercised in the cases of all institutions which are not under the Board of Superintendence. But in the case of the hospitals whose statistics are returned to Parliament by that Board, every confidence may be had as to their perfect accuracy. I fail to see why similar reports should not be issued concerning the county infirmaries of Ireland—institutions which are heavily subsidised, and about which very little trustworthy information can be obtained. But in the case of the six hospitals now under consideration, no question of the accuracy of their statistics can be entertained for a moment, and it is quite evident from these that one of three conclusions must be arrived at : either the Irish poor suffer from far less severe diseases in mid-life than is the case either in Scotland or England; or that the Irish hospitals admit much more trifling cases than do those of the sister isle; or that their hygienic conditions and the results of their practice are much better. To establish the first conclusion it would be necessary to have a mean age of the patients of every hospital, and the mean age at death of the three countries. The latter facts are at hand, but do not yield any evidence in favour of this first conclusion. The mean death age seems, as far as I can make out, to be rather lower in Dublin and Belfast than it is in English large towns, probably on account of very high infantile death rate. The second of my suppositions is one on which no statistics can be got to throw light, but having some experience of four out of six of these hospitals, I am not in a position to admit that their cases are less serious than our own. In favour of the third view, there stand the figures that the mean residence in these hospitals is 1·67 days below the average of all the hospitals returned; 4·43 below that of the London hospitals, 2·02 below that in the English "large town" hospitals, and higher than the residence in the cottage hospitals only. Then the general patient death rate is only 6·108, or very little more than the average, and better than that of any class of hospitals except those having less than 20 beds.

These facts are very striking, and are such as will not admit of any speculative explanation. It rests with the managers of the English hospitals and Scotch infirmaries to show something in the nature or conditions of their population such as will prevent all sanitary improvements reducing their death rate to something near the level of the cottage hospitals of England and the large hospitals of Ireland. In the case of the latter it must be borne in mind that they nearly all admit zymotic diseases; and in the case of one of them, the " House of Industry" hospitals, these cases amount to 26·7 of the total admissions.

Dr. Steevens's Hospital, with a mortality of only 2.8 per

cent., is practically the hospital of *dernier ressort* for the whole of the Irish constabulary, and therefore it admits a large number of serious cases. It has a marginal accommodation of nearly 43 per cent.

The Mater Misericordiæ has been at times populated by the victims of serious epidemics, and yet its mortality is only 7·7 per cent. The Meath Hospital has had, and still has, a staff whose reputation is as brilliant as that of any similar institution in Europe, and it also admits a considerable proportion of fever cases, yet its death rate is only 6·5.

There can be no question that the low mortality, and especially the diminished residence in the Dublin hospitals, must be to a considerable extent due to the repeated inspections and complete annual reports of the Board of Superintendence, and I think the time has arrived when every medical charity should be placed under similar supervision. To quote the words of Dr. Farr in the " Supplement to the Thirty-fifth Annual Report of the Registrar-General :" " What is wanted is a staff officer in every county, or great city, with clerks to enable him to analyse and publish the results of weekly returns of sickness to be procured from every district ; distinguishing, as the army surgeons do, the new cases, the recoveries, the deaths, reported weekly, and the remaining in the several hospitals, dispensaries, and workhouses ; these compiled on a uniform plan, when consolidated in the Metropolis, would be of national concern. It would be an invaluable contribution to therapeutics as well as to hygiene ; for it would enable the therapeutist to determine the *duration* and the *fatality* of all forms of disease under the several existing systems of treatment in the various sanitary and social conditions of the people. Illusions would be dispelled ; quackery, as completely as astrology, suppressed; a science of therapeutics created; suffering diminished; life shielded from many dangers."

I have placed together in another group twenty English county infirmaries, selecting those which had as little as possible the disturbing influence of a manufacturing population. They have an average marginal accommodation of 27 per cent., the mean residence in them is 40·24 days, and their average mortality is 5·217. In these figures there is ground for the suspicion that there is a want of vigilance in the management of these infirmaries. They are almost uniformly conducted on the system of admission by subscribers' tickets, and partake therefore more or less of the character of amateur workhouses. It is probable that their mortality might be reduced even below what it is by more careful management; for though it is constantly said of them that they seldom admit other than chronic cases, they are known to suffer, every now and then, from endemics of hospital diseases. If we contrast individual instances, we find that the first two on the list yield by comparison strong evidence of the advantage of a short mean residence. The general death rate of Leicester is 26 in 1000, and that of Exeter is 25, and the mortality of the Devon and Exeter Hospital is 1·21 per cent. less than that of the Leicester Infirmary. But in the latter institution the bed rate is double that of the Exeter Hospital, the mean residence being 18·14 days less. The Leicester reports show that the fever admissions are 5 per cent. of the whole, whilst the fever deaths constitute as much as 16 per cent. of the total mortality. On the other hand, the reports of the Devon and Exeter Hospital give no information. These figures make it probable that the hospital death rate at Leicester *might* be reduced, but they make it certain that at Exeter it *ought* to be less.

The highest mortality in any of these hospitals, 8·04 per cent., and the shortest mean residence, occurs at the Royal Bath Hospital. This is very remarkable, for Bath is hardly a town where we should expect a higher hospital death rate than in Leicester, Shrewsbury, Carlisle, or Norwich. Its general death rate is only 22 in 1000, and the margin of hospital accommodation is 37 per cent. The hospital is conducted on the ticket system, so that the short mean residence is very creditable to the management, especially when we also consider that about 56 per cent. of the cases are surgical. The reports issued by the committee are much more complete than such documents usually are, but they do not yield any information which can be made to explain the exceptionally high mortality.

The death rate in the North Devon Infirmary, at Barnstaple, is the lowest of all the large hospitals in Great Britain, but the residence is much above the average. The conditions which produce this extremely low death rate, if they could be obtained, would be most valuable in assisting us to draw general conclusions. There can be no doubt that the low district mortality, 18 in 1000, very materially assists in this desirable result.

That a considerable amount of an unusually high hospital death rate is, sometimes at least, due to intrinsic causes, is proved by the experience of the Norfolk and Norwich Hospital, which for 90 years had an almost constant death rate of 5·5 per cent. From 1861 to 1870 it was 5·656, the mean residence being 41·5 days. During the last six years the residence has risen to 42·3 days and the mortality to 7·7 per cent., both changes being due unquestionably to hospital influences. The average population has been reduced, and the annual number of patients has very materially fallen off; and the rate at which these changes have taken place seems to indicate that the confidence of the hospital constituency has been greatly shaken in the safety of residence within its walls. That there has been ground for this is shown by the admirable and honest tables in its reports; for in 1873, of eighteen deaths after operations, seven are returned as having occurred from pyæmia.

In a paper read before the British Medical Association in 1874, Dr. Beverley gives a full account of this rise in the mortality of the Norwich Hospital—a paper which may be consulted with advantage by all who are interested in hospital management. He says that rather than continue such a state of matters, "it would be better far to do away with the hospital entirely, and let those who now unconsciously run the gauntlet of its hidden dangers submit to surgical treatment in their own cottage homes, where they would have an undoubtedly better chance of recovering, even from the greatest accidents and operations, than in the wards of our hospital as it now exists, even with the aid of efficient nursing, good food, and the care and skill of its surgical staff." Dr. Beverley maintains that pyæmia is a disease produced exclusively by hospitals, and he quotes Mr. Cadge, one of the most distinguished of living provincial surgeons, to the effect:—"I have unwillingly and almost tremblingly proceeded to operate in the hospital; but I have had a happy confidence and a perfect assurance that in all private cases I should avoid any of those disas-

trous consequences, and I came to the conclusion in my
own mind that pyæmia, if it do not find its birthplace,
does find its natural home and resting-place in hospitals ;
and although a hospital may not be the mother of pyæmia,
it is its nurse." These are remarkable and strong expressions, but they seem justified by the facts ; and surely if we
can get such clear evidence of the influence of intrinsic
causes on the death rate of a comparatively small hospital
like that at Norwich, with a mortality of only 7·7 per cent.,
the disclosures which might be made by an equally careful
examination of the returns of hospitals with far higher
death rates, of which at present we have no detailed
account, would be sufficiently appalling.

Looking back to the great history which the Norwich
Hospital has had for more than a century, bearing in mind
that some of the most brilliant feats of surgery have been
performed there by men whose names are historic, and further
considering that its results are fully stated in its reports,
I am driven to the conclusion that if for 90 years it has
maintained a death rate of 5·5 per cent., it is incumbent
on every general hospital whose mortality may be much
higher to render an account of the same.*

As great stress has been laid upon accident cases as a
cause of hospital mortality, especially in the London hospitals, I have taken out the details of a number of hospitals
devoted exclusively to the reception of accidents. They

---

* The history of this hospital is also suggestive of the conclusion I have
already indicated, that the triumph of medicine lies in prevention, and that its
efforts towards cure have not as yet exercised any tangible influence. With
the single exception of ovariotomy, it cannot be shown statistically that our
curative efforts have resulted in any marked prolongation of human life. Even
in ovariotomy our success seems owing chiefly to preventive measures directed
against septic infection.

are nine in number, that being all which are available for
my purpose; and, very unfortunately, I am unable to
include amongst them the returns of the accident hospital
at Poplar. I have them, however, for the preceding decade,
so that they may be very fairly used in comparison.

The class of accidents received into these hospitals is
peculiarly severe. I have had some years' experience of
one of them, and am in a position to assert that the smashes
and burns admitted could not be surpassed in severity by
what is seen in any large hospital. Their mortality is
therefore necessarily high, but yet it is less by ·2 per cent.
than that of all the London hospitals; and their mean
residence is only slightly greater—a fact which is due to
the existence in them of a disproportionately large number of
cases of burn, the residence of which often extends over many
months. The general mean residence would be greatly
less than that of the London hospitals if the return of the
Launceston Hospital, where it amounts to fifty-four days,
were excluded. I think it very likely that careful attention
to segregation might very much diminish their mortality;
for I have often seen pyæmia prove fatal to accident cases
apparently from the presence in the same ward of a case
of extensive suppuration from a burn. In all hospitals, but
especially in those small accident hospitals, every such case
should be carefully isolated.

The returns from these accident hospitals seem to show
that in the London hospitals, and in others where the mortality is as high as ten or twelve per cent., the accidents
are not sufficient to account for the excess.

I have placed the Irish county infirmaries in a group
by themselves; for it seems as if, for some reason or other
of which I have failed to get a satisfactory account, they

cannot be fairly compared with any other kind of hospital. There are 28 of them, and for the years from
1861 to 1870 I obtained returns from 18, but only nine
have replied to my last circular. The reports which they
issue—at least those which I have seen—are very deficient
in information. Of the 19 from which I have had no
replies, six have Government grants amounting to 487*l.* per
annum, the proper expenditure of which does not seem to
be under the control of any central body, as is the case
with the Dublin hospitals. Twenty-four out of the whole
number are subsidised by county grants, amounting to
20,104*l.* for last year ; and for this large amount of money,
contributed from the public rates, it is surely not too much
to expect that satisfactory accounts should be rendered, both
in financial and medical details. The number of in-patients
treated by these 28 hospitals during last year amounted to
only 11,974—that is, they had 33*s.* 6*d.*, or nearly 13½*d.* a day
for every patient treated. From 1861 to 1870 the average
mortality of the 18 infirmaries from which returns were
obtained was 1·95, and the mean residence was 29·85 days.
Of the nine returns obtained for the last six years, the
average mortality is 2·726, and the mean residence only
25·27 days.

These figures are very difficult to understand, and though
I have made numerous inquiries amongst the officials of
these institutions, and others likely to be acquainted with
them, I have received no information which helps me to
explain how these county infirmaries can conduct the
treatment of such diseases as must necessarily exist amongst
the Irish peasantry, and yet have such low death rates.
Either these institutions deserve to be taken as models of
hospital salubrity, or they must be doing to a great extent

the work of relieving-officers ; and in either case full information is needed concerning them. In Howard's time,
these institutions were greatly in need of inspection, and
I have failed to find that they have it now. His descriptions of what he saw in some of them are really terrible.
Generally he found them in an unfavourable state, the cost
being greatly out of proportion to the patients treated.
The quantity and quality of the linen was almost always
defective, the floors apt to be sanded to hide the dirt, and
the patients sometimes bedded in close boxes. The diet he
describes as deficient, and that this fault exists still is seen
by the diet-sheet published in the annual report of the
Louth County Infirmary. It contains a regimen much
more resembling that fit for the punishment of prisoners
than for sick people. The cost of the diet per patient for
each day is given at 6*d.*, yet the infirmary has Government and county grants to the amount of 713*l.* per annum,
which is an allowance of 52*s.* 7*d.* per patient treated, or
rather more than 2*s.* a day. The total income of the hospital is 1076*l.* a year, which gives an expenditure of nearly
4*l.* per patient, or 3*s.* a day. It remains a very interesting question of public economy how the 3*s.* a day is
spent on each patient, if his diet costs only 6*d.* A
consideration of these details rather inclines me to the
belief that there is more of the workhouse than of the truly
hospital element in the Irish county infirmaries—a belief
which is strengthened by the small amount of surgical work
which is done within them. In Howard's " Account of
the Principal Lazarettos in Europe," page 83, he tells us
that in one of these county infirmaries 2*d.* a day was
allowed for the diet of each patient, and he very pertinently
asked the governors to consider that criminals in the county

gaol had 3*d.* a day allowed for their diet by Act of
Parliament. The diet-sheet in the Louth County Infirmary
seems to be pretty much what it was a century ago.

One of the most interesting groups of hospitals undoubtedly would be that containing the institutions devoted
to the treatment of children, if complete and satisfactory
returns could have been obtained from all of them. For
the years from 1861 to 1870 I had returns from 12 of
these hospitals, but to my last circular only six replied.
These are both too few and incomplete to make any perfect
deduction from, but they make it quite evident that these
various institutions are conducted upon very different principles.

The largest and most important hospital for children
in England is that in Great Ormond Street. In the first
return the mean residence during the decade was 37·2 days—
a period which, I think, must be unnecessarily protracted
if we bear in mind the rapid course which diseases usually
run in children. The building occupied was very ill-
adapted for the purposes of a hospital, and I think we may
conclude that the mortality of 11·421 per cent. was considerably higher than it would have been under more favourable circumstances. This view is supported by the evidence
of the hospitals at Edinburgh and Birmingham. The
former of these does not admit surgical cases, and is therefore likely to have a relatively high mortality, though it
only reached 10·347 per cent., with a mean residence of
39 days. In the Birmingham Hospital for Children the
mortality was only 7·359, and the mean residence was
24·2 days, facts which are sufficient alone to indicate this
hospital as a thoroughly well-managed institution. Amongst

its rules is one that all other hospitals should adopt—to the
effect that no patient is allowed to remain in the hospital
more than a month without the sanction of a general
consultation.

When the mortality of a children's hospital is found
to be so low as 2·3 per cent., I think it may be fairly
inferred that such an institution is not doing the work it
ought to do ; for such a mortality can only be arrived at
by the exclusion of acute cases.

Of the four returns obtained for the last six years, the
average mortality is 6·12 per cent., and the mean residence
is 24·72 days. The mortality of the Ormond Street Hospital has fallen to 10 per cent., and that of Birmingham to
7·14, which is the same as that of the Victoria Hospital in
London ; whilst the mean residence in the latter is 16·42
days higher than it is in Birmingham. The institution in
Ormond Street has recently been transferred to a mag-
nificent new building, where it is to be hoped better results
will be obtained. At the Birmingham Hospital, isolated
wards have recently been built for the treatment of zymotic
diseases, and for those . scourges of childhood, croup and
diphtheria. There is also to be a quarantine ward, and this
example should be copied by all hospitals for children. The
tables published in the reports of this hospital are such as
may serve as models for every similar institution.

Eight returns have been obtained from hospitals devoted
to the treatment of diseases peculiar to women, but the
results are so unequal as to be quite unfit for any purposes
of comparison. These hospitals are essentially for the
treatment of chronic cases, in the majority of which there is
little or no risk of life ; and it is only when there is a

special activity on the part of the staff in undertaking the performance of certain formidable operations, that the mortality ever approaches that of any general hospital. Thus at the Samaritan and Soho Square Hospitals a large number of ovariotomies and kindred operations are performed, the fatal cases of which greatly contribute to bring the death rate up to nearly 6 per cent. On the other hand, it is quite evident that at the Leeds Hospital for Women, and at the Marylebone Road Hospital, such operations are never, or at least very rarely, performed. At the Birmingham Hospital these operations form a large proportion of the comparatively small number of in-patients, and the mortality is 8·1 per cent. In a hospital reserved exclusively for these cases the mortality might be as high as 25 per cent.

The hospital at Cork is a Government Lock Hospital, and therefore very rarely has a death.

LYING-IN HOSPITALS.

There is certainly no kind of hospital which has yet had so searching an inquiry into the reason of its existence and the results obtained by its use as the building devoted to the treatment of women in childbirth. So far, the verdict has gone almost entirely against lying-in hospitals, and on the whole this is not a matter of regret. The terrible experience at King's College Hospital forbids for ever the possibility of attempting to accommodate parturient women in the same building with other patients. But whether or not hospitals entirely for lying-in women may be conducted with less risk, or with the same amount of risk, as will attend women confined in their own homes, is a question which is not yet, in my opinion, fully decided. That most of these institutions have had unfortunate results is not conclusive evidence that these are inevitable. If in order to relieve human suffering and to help human poverty it is necessary to have hospitals of any kind, then hospitals for parturient women are as necessary as any others, and there remains only the need of discovering how they can be constructed and managed so as to have as much safety as if the women were confined in their own homes. It is quite as good an argument against military hospitals to point to the awful disclosures of Howard, as it is against lying-in hospitals to point to the high rates of mortality given by Le Fort and others. Utter demolition is not reform.

In order to determine what the mortality of lying-in hospitals should not exceed, it is first necessary to establish what is the average mortality of women in childbed in those classes of the population from which the inhabitants of these institutions would be, or ought to be, drawn.

This has never been done. We are occasionally favoured by reports of charities which conduct their practice at the houses of the patients; but on careful examination it is always found that their numbers are too small for any just deduction; and, secondly, that they have always some conditions attached to them which introduce such source of fallacy that no mere extension of numbers would remove it. In dealing with general illness and accidents the use of large numbers seems likely, though not absolutely certain, to remove error. But in a case where the conditions are constant, as in parturition, the increase of the numbers will probably only increase the extent of the error.

Thus these maternity charities are frequently found to confine their efforts to "poor married women," a condition which at once removes one of the great sources of parturient mortality—the treatment of unmarried, seduced, and deserted women. Again, first pregnancies, even among married women, are greatly diminished; for it is found that in their first trial women are generally attended by an accoucheur, the young couple being generally in a position to pay some sort of fee.

To provide gratuitous attendance for married women seems to me a mere encouragement of improvidence, and a method of charity which ought to be at once discontinued. Those who really want help most are those who, by our present plan, have least chance of getting it. The women amongst whom puerperal mortality is always found to be highest, are those we often send to be confined in our workhouse wards. I do not know that the mortality amongst unmarried primiparæ in these institutions has ever been properly displayed, but I think there is reason to believe that it would be better for all concerned, if we had these cases treated by themselves, in buildings set apart for the purpose, at least in our large towns. I think it is a question worth considering whether the existence of institutions where unfortunate girls could be readily and at once admitted, in the hour of their trial, would not greatly diminish the number of those terrible child murders which occupy so much of the time of our coroners' juries. The objection that such institutions might have a tendency to encourage vice is no answer, unless it can be shown that our present neglect diminishes it—a supposition which is highly improbable.

The existence of such hospitals would stand to parturient women precisely as our recently established hospitals for zymotic diseases do to our general hospital population. They would remove from their midst the greatest and most constant source of danger.

For the discussion, therefore, of the mortality of lying-in hospitals it seems to me that we are not yet in possession of the proper data, and I cannot say that I see where they are to be obtained. They never will be until some system can be devised and put into operation for the accurate record of all hospital statistics. I must say that I have failed to find any set of obstetric records which are such as bear the impress of so great a degree of exactness as to be infallible as a basis for comparison. This is to be the more regretted as it is perfectly evident that what could be predicated against lying-in hospitals might be inferred against all others. At present it is the fashion to express satisfaction that they have been or are likely to be disestablished; but it is forgotten that if this policy is necessary for them, it may legitimately be advanced against almost every other kind of hospital.

I have obtained returns from ten lying-in hospitals, which
have treated an aggregate of over 22,000 cases in six years,
with an average mortality of 1·061 per cent. But that tells
me nothing more than the fact of so many women having
died during or after labour. Seven of these hospitals give
returns so complete that I am able to see that three are
managed on principles totally different from those which
regulate the other four. In these three the average residence is over twenty days, so that the women are probably
selected cases, admitted some time before labour sets in, yet
the mortality is 1·377. In the other four the mean residence is 9·81 days, so that the cases are evidently all admitted in emergency, and the mortality is only 1·0005. In
these two classes the mere difference in the mortalities—
·3765, by no means represents the real contrast between
them, for amongst the four is to be reckoned the Rotunda
Hospital of Dublin, which alone admits during the year 30
per cent. more than all three of the other class put together,
and includes amongst its patients a large number of unmarried women, amongst whom the mortality is much higher
than amongst the married. I doubt if any other of the three
hospitals of the first class admits unmarried women. It is
very remarkable, and should be made a matter of strict
inquiry, that in the hospital in Endell Street, where the
residence is most prolonged, the mortality is also the highest.
That unmarried women and primiparæ are more subject
to puerperal disease is so fully established that it should be
made equally clear that their hospital conditions should be
made special.

Statistics of twenty-seven London workhouses, with a
mortality of ·62 per cent. of the women confined there, are
given by Miss Nightingale, together with a statement of the

experience in the Liverpool Workhouse lying-in wards, with a mortality of ·56 per cent. These, though the figures are comparatively small, would have been of great value if she had also given the number of primiparæ and their position, whether married or single. Such institutions have a large number of births amongst women technically unmarried, but living in a state of concubinage, which, in the great majority of cases, is quite as faithfully maintained as in truly matrimonial life amongst the same class.

It is, therefore, the number of unmarried primiparæ, and the mortality amongst them, which is the information wanted for all institutions.

The figures of the Rotunda Hospital, published in various reports of that hospital, and very carefully considered by Dr. Matthews Duncan in his book on the " Mortality of Childbed and Maternity Hospitals," are beyond doubt the best record of obstetric cases we can obtain. But even with them the bare statement of the number of deaths and the number of patients treated during each year, will yield no very definite conclusion. Dr. Duncan splits these figures up into groups with a view of showing that mere aggregation has no influence on hospital mortality. He puts together the figures of various years according to the population of the hospital, or the number of women delivered. But he does not regard the fact that it is quite as easy to crowd ten women as it is to crowd a hundred. The Rotunda Hospital is, like all large hospitals, a series of small buildings stuck end to end and put on the top of one another, each of which may have a different mortality. We know how fond hospital managers are of closing wards at times when it would be better that they should be open ; and it is within my experience to have seen a hospital most crowded

when its inhabitants were really fewest. Mere increase of
population does not necessarily mean increased density, and,
therefore, unhealthy overcrowding. Dr. Duncan finds that
at times when he gives the mean age of the hospital building
to be 49 years, the mortality runs from ·848 to 7·25 per cent.,
and in other periods, the mean data of which is 1838, the
mortality varied from ·58 to 3·82. But he gives no evidence
that to account for the higher mortality of these periods
there may not have been some terrible overcrowding, localised
in some ward or wards, or extending over such a brief period
as to leave no mark on the general hospital population.
Besides, to give the figures absolute value they ought to be
compared with the outside puerperal mortality amongst the
hospital constituency, for the corresponding periods.

As a conclusion from this table, Dr. Duncan says, in
italics, that " the mortality of the Dublin Lying-in
Hospital *does not increase with the increased number of
inmates—does not rise with the aggregation."* This may
be granted, without its affecting in any way the general
belief that the mortality is greatly affected by overcrowding, for it might be just as well argued that because
the enormous increase of the population of London does
not bring with it an increased death rate, that aggregation
of the population, such as it is seen in Seven Dials, the fever
dens of Liverpool, or the West Port and Cowgate of Edinburgh, may occur with safety. It must be always borne
in mind that what seems to be true of all hospitals must be
eminently true of lying-in hospitals, the chances of the occurrence of septic centres being not only increased with the
increased size of the hospitals, but the chances of their
being originated by inattention to sanitary requirements, and
of their finding a soil suitable for their propagation, rise in

a greatly increasing ratio. This is really the secret of the success of small hospitals over large ones, which seems to be established by my general tables.

Large hospitals are not more unhealthy because of their greater size, but because they want more looking after, larger bed areas and cubic spaces—conditions which they do not obtain because in places where they exist, time, labour and space are so much more valuable than in the small hospital areas. But it must never be forgotten that a small hospital may be made as unhealthy as a large one. Small size is no guarantee of salubrity. Of this Dr. Duncan gives convincing proof in quoting the statistics of the Edinburgh Maternity. This was a small old-fashioned confined house, as unsuited for a hospital of any kind as it is possible to imagine. That many women came out of it alive is really a matter for congratulation. Now, I am told, no register of the cases is kept, and at least there is probably good reason for its discontinuance; for as Dr. Churchill, in the *Dublin Quarterly Journal of Medicine* for 1869, gives the returns of this hospital for the years from 1844 to 1868 with a mortality of 1·64, we of course know that a register was at one time kept. To pick out a hospital like this and say that small hospitals are no better than big ones is of course a fallacy. They *can* be made quite as bad, but it appears more difficult to make them so.

From a very valuable table (No. XXV.) in his book, Dr. Duncan concludes that about one in every 2·2 cases of puerperal death is from metria. But he takes for his standard of comparison the statistics of private practice, the mortality of childbed in towns of Scotland, those of some eminent obstetricians in Dublin, and those of the hospitals in Ireland. It is perfectly evident, however, that there is

a great source of error in not dealing with known quantities of primiparity. This is such a constant source of
danger that no conclusions of childbed mortality can be
secure when it is not known. If the puerperal deaths
were found, say in Dublin, to be higher in the hospital than
they were outside, after correction for primiparity and the
unmarried, then the Rotunda Hospital should be closed at
once, and with it all other maternities, if it were also found
that this was an inevitable result of congregating a number
of lying-in women together. As yet this result is not known
to be inevitable, and therefore the wholesale condemnation
which lying-in hospitals have received is premature.

In the 17th Annual Report of the Registrar-General, a
table is given by Dr. Farr from which Dr. Duncan extracts
the following :—

Ages.	No. of Child-bear- ing Women.	Deaths from Puerperal Fever.	Mortality per cent.
15—24 ...	107,440	... 298	... ·277
25—34 ...	328,720	... 486	... ·148
35—44 ...	166,140	... 256	... ·154
45—54 ...	7,545	... 12	... ·163
Total	609,845	1052	·172

It cannot of course be pretended that the value of the
percentage is absolute in such a table. It is open to the
objection that all such tables are liable to—that of incomplete returns. But its value is great as establishing a ratio
of puerperal fever mortality for different periods of life;
and the largeness of the numbers used makes it almost certain that this ratio will be constant. We have it then
established that between the ages of 15 and 24, when of
course the great majority of first labours occur, that the

puerperal fever mortality is ·105 per cent. above the average. But this is subject to the correction, for our purpose, that all the cases are not first labours in this age, and that probably there was a large proportion of the deaths in the other periods which were primiparæ. Hugenberger's tables on this subject, also given by Duncan, are of very little use, for the total mortality is so bad that probably the ratio between the primiparous and multiparous deaths is greatly disturbed. He gives the primiparous puerperal-fever death rate at 4·31, and that of multiparæ as 2·4, both of which are, I trust, exaggerated by intrinsic hospital causes. Dr. Duncan has, with infinite labour, compiled a table from the returns of Edinburgh and Glasgow for 1855, which yield 19,104 cases. Amongst these the primiparæ were to the multiparæ as 19·5 is to 80·5. This would have been a valuable constant if the figures used had been large enough, but they are not. Still less valuable is the statement of the mortality, which in primiparæ was ·698 per cent., and in multiparæ rather less than half—·338 ; for it is evident that it could be used only as a standard of comparison for hospitals in the towns of Edinburgh and Glasgow. It is useful, however, as tending to corroborate conclusions from the other sources. From the habits of the people, engendered by the peculiarities of race, religion, and the state of the marriage laws in the two countries, it is quite impossible to compare the puerperal fever mortality of Scotch towns with that of Irish hospitals. The "murderously depressing influence of shame," to use Dr. Duncan's forcible language, is not felt by unmarried women in Scotland as it is in Ireland.

In Hardy and McClintock's " Midwifery and Puerperal Diseases," 9852 cases are tabulated from the practice of the

Rotunda Hospital, 31 per cent. of which were primiparæ, with a mortality of ·4 per cent., whilst the death rate of the multiparæ was only ·22. But none of these small collections is enough to give a constant ratio, though probably we may accept Duncan's conclusion that both the general and puerperal fever death rate of primiparæ is double that of all multiparæ. Of the primiparæ, those who are unmarried present in all probability a much higher death rate than those who are married. I have not been able to obtain material in large enough amount to put this conclusion in figures, but I hardly think it will be disputed. Dr. McClintock tells in his returns of the practice of the Rotunda during his mastership there (*Dublin Medical Journal,* vol. xiii. p. 272) that 127 patients were unmarried, and that of these 31, or nearly one-fourth, died, chiefly from some form of metria. This experience is substantiated by Dr. Johnston. We may conclude, therefore, that if we had a hospital exclusively for unmarried primiparæ we might possibly have a mortality as high as 20 per cent. Anything below that in Dublin would certainly be a gain, and it really seems to me that to establish such a hospital in some of our large towns would be an experiment worth trying. I am not sure that it would not be the only maternity hospital or charity which we should be justified in maintaining. If a woman falls into trouble a second time, her risk is very much less and her guilt greater; but from a large experience of unmarried primiparæ in hospital practice, I am quite certain that they are more sinned against than sinning; and I am convinced that they ought to be specially protected from the risks they have incurred by their fault, both for their own sakes and for the sake of other women to whom metria may be communicated from them.

A great many papers have been written on lying-in hospitals, but they are chiefly directed as attacks on, or defences of, special institutions, and do not shed much light on the whole question.

Looking at the returns which I have been able to obtain, I can only conclude that if Dr. Farr's estimate of ·483 per cent. mortality is to be accepted as including both deaths from metria and accidents of childbed, as I think it is, as the constant death rate of all labours, the death rate of some of these institutions, which admit only married women, is far too high and demands an explanation.

In Miss Nightingale's " Notes on Lying-in Institutions," that distinguished authoress speaks of labour as " not a diseased, but an entirely natural condition" (page 10), and thereby she has perpetuated an error evident to every gynæcologist. We do not know the percentage of all cases in which pelvic deformity requires instrumental interference, but we know that they are on the increase, induced unquestionably by our altered habits. In each of these cases labour is a diseased process. The very fatality following the labours of unmarried primiparæ is a result of civilisation, and one, therefore, which society is bound to provide against as far as possible.

It is impossible in the space at my disposal to follow Miss Nightingale through the array of statistics which she brings to bear against lying-in hospitals, but my general conclusion is that they are but little to the point. To tell us by a table, the data of which need not, for my purpose, be disputed, that the mortality of women confined in the Paris hospitals is nearly 8 per cent., or that other hospitals approach this more or less nearly, is to tell us what every one will admit,—that puerperal women are specially

prone to be affected by any contagion near them, and
no one will now dispute the utter impropriety of treat-
ing puerperal women in general hospitals. But when,
summarising her conclusions, she tells us that, making
allowance for inaccuracies, there is a higher death rate
in lying-in wards than in home deliveries, and that the
great cause of the excess is blood-poisoning, her deductions
are manifestly open to correction, if they mean that lying-
in hospitals should no longer exist because a high death
rate in them is inevitable. We do not yet know the constant influence even of the most fatal condition of parturient
women, and we have by no means exhausted all possible
attempts to make these hospitals safe. The great majority
of women do not require them, but there is an important
minority which must continue to exist as long as human
instinct remains ; and the question is really whether or not
these unfortunates should be specially cared for, and whether
they had not better be cared for in hospitals. For married
women, save in rare cases where hazardous operations have
to be performed, lying-in hospitals are, in my opinion, not
only not needed, but even home practising charities should
be greatly discouraged, unless they can be conducted on the
provident principle. In all hospitals we must have two
objects constantly in view—to relieve human suffering with
the minimum cost of life, and also with the minimum
tendency to diminish the self-reliance which is so easily
knocked off its pedestal amongst the classes from which our
hospital population is drawn,

The last of the groups into which I have divided my
returns includes all the information I have been able to
obtain concerning hospital accommodation for zymotic

diseases. Special fever hospitals are of comparatively recent origin, and are still greatly wanted in many places where zymotic diseases prevail. They constitute a class of hospitals against which I think no objections can be raised, chiefly because it is to be hoped that they are only a temporary expedient.

We do not yet know with absolute certainty that vaccination, however perfect, will completely stamp out smallpox, but the evidence is conclusive that if every human being were once satisfactorily placed under vaccine protection, the disease would cease to have the mortality of nearly 20 per cent., which these returns show.

It would appear from the facts which I have been able to gather concerning the smallpox epidemics of the last six years, that a wave of this disease has recently passed over the whole country. The differences seen in the mortality of smallpox in various towns are probably in greater part due to a difference in the extent of the protective influence of vaccination, but there is reason to suspect that here, as elsewhere, hospital conditions interfere to produce a higher mortality. Both of these influences are probably in operation to cause the difference in the smallpox mortality of Greenock and Paisley. It is more than likely that the larger floating population of Greenock brings into that town an undue share of smallpox nidus; but we have already seen that there is a greater tendency to death generally amongst the patients at the Greenock Infirmary than there is in that at Paisley.

The long roll of hospitals which have to admit typhus fever, and the terrible mortality it inflicts, are the result of a neglect of sanitary precautions discreditable to our advanced civilisation. This disease is one which is entirely removable, yet there are few towns of any size in which it does not occur, and it has an average mortality of nearly 16 per cent. It is inexpressibly shocking to find that in London its mortality is so high as 21·5 ; and nothing could be more convincing of the need there is for legislative control over the building and management of the houses erected for our working population than a bare statement of the facts of this disease.

Enteric fever is another disease which we have reason to think might be completely stamped out by sanitary improvements, yet there is probably no other fever which has been of late so chronically endemic in both town and country population. For the first time we have been able to gather an idea of what its real mortality may be from the returns of the Homerton and Stockwell Hospitals, where 18 per cent. of the cases have died. This, of course, may not represent the absolute mortality of the disease ; for, from its peculiar character, it is likely that only the more severe cases are sent to these hospitals. In this respect it differs from -smallpox, and perhaps also, though to a less extent, from typhus.

So little is known with certainty of the origin of scarlet fever, and in the two separate returns obtained of this disease there is such a striking difference in the mortality, that nothing very definite can be said about it.

Some very curious information is to be obtained from the returns issued by the Metropolitan Asylums District Board. Thus, if we take their statistics for the years from 1872 to 1875 as indicating the zymotic condition of the metropolis, we may conclude that scarlet fever was at its ebb in 1873, and that then its mortality was lowest; and that when it was epidemic in 1875 its mortality was highest, as might be expected would be the case with all zymotic diseases. But this does not seem to hold good of enteric fever, for when most prevalent its mortality was lowest, and it rose when the admissions were diminished. This is still more markedly the case in typhus fever; for when clearly epidemic in 1874 its hospital mortality was lowest, and in 1875, when the admissions were only 11·9 per cent. of those of the previous year, the mortality rose. Taking all the zymotics of 1874, we find that the hospital mortality was 42·8 per cent. less than in 1872, during which year the admissions were not 30 per cent. of those in 1874. The mortality during 1874 was also 1·2 per cent. less than it was in 1875, though during the latter year the admissions were 12·5 less than those in 1874.

The recent establishment all over the country of medical officers of health will, it is to be hoped, when worked upon a more uniformly complete plan than it is at present, contribute to a knowledge of the natural history of zymotic diseases, such as will enable us to perfect our measures for their suppression. The general facts seem to show that, after smallpox, typhus is the most fatal of the zymotics; that enteric fever ranks next, and then scarlet fever—an arrangement which is somewhat unexpected.

A general study of the zymotic returns seems to give an approximate value to the statements of the patient rate, the bed rate, and the mean residence of hospitals generally. Thus it appears that a diminished bed rate concurrent with a high patient rate means either prolonged residence or the statement of a larger number of beds as in constant occupation than really is the case. If the former, it may indicate an undue preponderance of chronic or surgical cases;

or, in the zymotic hospitals, of a large number of cases of enteric fever. It may also mean bad sanitary arrangements in any hospital. A relatively high bed rate with an average or low patient rate means rapid recovery or trifling cases. When both the patient rate and the bed rate are low and the residence is prolonged, it may be suspected that the usefulness of the hospital would be greatly increased by better administration. When both patient rate and bed rate are high, and the residence comparatively short, there is reason to believe that some intrinsic defects exist in the hospital. Finally, a high patient rate should always be regarded as calling for constant care on the part of the executive to guard against overcrowding, no matter what may be the character of the disease or diseases treated within the hospital. Its diminution in relation to a constant bed rate would mean diminished residence; and if we could suppose the characters of the constituency to remain the same, any such diminution would infallibly mean greater success for the work of the hospital.

The determination of the causes of excessive mortality in hospitals of any kind is a most difficult task, for we have first of all to contend with the almost insuperable difficulty that there are very few data which can be depended upon as entirely free from error. Mere expressions of opinion, as I have already said, are of little value, however weighty may be the authority from which they emanate.

It is greatly to be regretted that the elaborate report written by Dr. Bristowe and Mr. Timothy Holmes, and issued in a Blue-book for 1864, is so open to this objection. It is very deficient in conclusive facts, for its statements are given for various years upon no uniform plan, and some hospitals are chosen for comparison and others neglected without apparent reason. General and somewhat vague impressions are made to do duty for facts, as is shown by a passage at page 509, where it is stated that " the mortality of the small and large London hospitals does not vary, or if it does the small hospitals have the larger mortality." The returns of the London hospitals for the last six years, as given in my tables, are quite enough to show the fallacy of this general impression; and they are abundantly confirmed by the less complete returns for the previous decade. On the same page it is also stated that at the Dover Hospital there is a low mortality, because nothing but chronic cases and a very few accidents are admitted; whilst at the Hemel-Hempstead Hospital the surgical practice is more active and therefore the death rate is higher. But the facts are just the reverse. The mortality of the Dover Hospital for the last six years has been 5·42 per cent., or ·248 above the average mortality of its class; whilst the Hemel-Hempstead Infirmary has, during the same period, had a mortality of only 3·33 per cent.—that is, 3·276 below the average mortality of similarly-sized institutions. The mean residence at Dover is nearly four and a half days less than it is at Hemel-Hempstead. During the ten years from 1861 to 1870, which includes the very time for which the report in question was made, the Dover mortality was 5·218, and that of Hemel-Hempstead was 3·811; so that over a period of 15 years the mortality of the two hospitals is found to be singularly constant.

At page 536 the report says—" It may be stated generally that patients remain longer in country than in town hospitals." No facts are given in support of this, and the statement is shown to be erroneous by the results of my circulars. The mean residence in the London hospitals is 32·91 days, that in the "large town hospitals" is 29·5 days, and that in purely country hospitals, even including the mismanaged county infirmaries, is 29·2 days; whilst in the purely rural or cottage hospitals we have a mean residence of only 24 days.

I have selected this report for the above criticism, because it is the most important document which has yet been issued upon the hospital question, but similar objections could be raised against many other contributions on the same subject. But I think I have said enough to establish a position which could hardly be at any time disputed.

Dr. Bristowe and Mr. Holmes tell us that they "have been led irresistibly to the conclusion that the chief cause of all the differences, real and apparent, which exist between different hospitals, is to be found in the constitution of the hospital itself." This is a conclusion which will find few opponents, provided it be understood that the word *constitution* includes the whole economy of the hospital, its management, and its hygiene.

There are certain features in the management· of a hospital which must of necessity influence its death rate, and first of these stands the nature of the cases for which it is intended. Nothing would be gained by contrasting the death rate of such a special institution as the hospital for consumptive patients at Bournemouth, 1·215 per cent., with the high mortality of the Greenock Infirmary. But hospitals of the same kind may fairly be compared, and it lies with them to give an explanation of any excess in their death rate. The admission of certain cases may, from the facts I have already given, be recognised as exercising an influence in the mortality, and I have shown that the admission of zymotic cases probably influences not only the general death rate of any hospital into which they may enter, but that it also increases the mortality of the non-zymotic patients. Dr. Bristowe and Mr. Holmes tell us that many large hospitals—notably those of University College, Charing Cross, and the Royal Free Hospital—exclude all cases of fever by special rule. This also means that they exclude a group of cases as large, or even perhaps larger, of such kind as may be mistaken for zymotics in their earlier stages. We should expect, therefore, that these three hospitals, being similarly situated, and doing very much the same kind of work, should have death rates nearly equal, and that they certainly should exhibit fewer deaths than hospitals known to admit large numbers of the severest zymotic diseases. But here we are disappointed, for University College Hospital has a mortality nearly double that of Charing Cross, and 4 per cent. higher than that of the Paisley Infirmary, where a third of the whole hospital population is affected by zymotic diseases. In the Blue-book report the writers select a few hospitals, taking them, as they say, almost at random, to illustrate " how utterly absurd and childish it is to compare hospital death rates without taking this element into consideration ;" but taking the whole of my statistics, and not selecting them at random, I find that the admission of zymotic cases is probably not nearly so important an element as Dr. Bristowe and Mr. Holmes have supposed. At least there must be some others of which they have not given illustrations, for I find that my chances of life would be about equal if I went into the Dundee Infirmary and had typhus fever, or was admitted into the splendid new hospital on the Albert Embankment with any kind of disease whatever, such as is usually treated there. It is a very grave question of social economy whether such a state of matters is inevitable.

I have been able to extract from the reports of the Manchester Infirmary for the years from 1861 to 1869 very complete information of the various causes of death. The mortality of the fever cases admitted were 18 per cent., and the fever mortality constituted 15·5 per cent. of the total. During the Lancashire famine years, singularly enough, the fever cases were below the average, and the mortality was only 16·6 per cent. They were most numerous in 1869, the increase being 96 per cent., and their mortality was 19·5. During the same year the medical admissions greatly exceeded their usual proportions, but all the cases of phthisis and tubercular diseases formed only 4·2 per cent. of the total admissions, and this is the highest ratio they presented. From the persistency with which general hospitals as a rule either refuse admission to cases of consumption, or get quit of them as soon as they can if likely to prove fatal, I do not think that this disease greatly influences hospital death rates, though Dr. Steele seems to think that its effects are sufficient to account for most of the differences observed in the mortality of various hospitals. In only one hospital I have found that it does so to any marked extent, and this is no doubt due to the fact that this particular institution—the German Hospital at Dalston—is really a refuge for foreigners in all kinds of illness, and that, therefore, all German patients suffering from phthisis in an advanced stage would probably be retained till death. Cases of phthisis constitute fully 10 per cent. of its entire population, and exercise a very marked influence on its mortality, and yet it contrasts favourably with institutions which do not admit nearly so large a proportion of this fatal disease.

Another element which has been supposed by Bristowe and Holmes to increase the relative mortality of the London hospitals is, that "the beds in London hospitals are allotted to medicine and surgery in proportions quite different from those in which such cases occur in actual practice. This preponderance of surgery in English, as distinguished from Scotch hospitals, is one of the chief causes affecting their sanitary condition." If this be true it is of course a subject for remedy as much as any other hospital abuse, unless it can be shown that the London hospital constituency suffers more from surgical diseases than from medical. They seem also to ignore the well-known fact that the mortality on the medical side of a hospital is always much higher than that on the surgical; and if the alleged preponderance of surgical cases does exist, the relative mortality of the London hospitals should be lower than that of the Scotch. I have already shown, however, that it is not so.

But they tell us, further, that this disproportion does not exist in the London, University College, or King's College Hospitals (Report, p. 468).

It is difficult, *à priori*, to see how this difference in the allotment of beds can make much impression when a large number of hospitals is dealt with, for human suffering and human diseases must be very much, if not wholly, the same under similar circumstances all over the country. We also find that the same disease is in one hospital classed as medical, whilst in another it may be placed under the care of a surgeon.

The diseases which are usually classed as medical, besides

H

increasing the mortality, always raise the bed rate and
diminish the mean residence. If in this light we compare
the Scotch infirmaries with the London hospitals, we find a
difference of 1·82 days in the mean residence, which, according to the report of Bristowe and Holmes, would, and probably does, indicate for the Scotch infirmaries a preponvera-
ting influence of medical cases, amongst which, as I have
already shown, is included a very large proportion of zymotic
diseases ; yet their mortality is ·351 per cent. less than that
of the London hospitals. This is conclusive proof that
mere disproportion between surgical and medical beds,
though it may have some influence, at least gives us no
tangible expression of it ; and it might be still further
shown that the better results which are obtained at St.
Bartholomew's Hospital do not appear to be due to any
special division of the beds between the medical and surgical
officers.

This question, however, leads up to one of far more
importance in hospital management, and which has been
greatly if not altogether overlooked in many institutions.
With a persistent conservatism which is worthy of a much
better object, the old-fashioned distinction between the phy-
sician and the surgeon is still kept up ; and, what is of far
greater consequence, because we retain this piece of anti-
quity we generally divide our hospital patients into the two
classes of surgical and medical, and place them accordingly
in different wards or buildings.

I presume no one will dispute that the dangers of overcrowding are infinitely greater for surgical cases than they are
for medical. At page 477 of Bristowe and Holmes' report
we have this opinion expressed as follows: " The exhalations
from a large number of acutely suppurating sores produce an

atmosphere in the ward which, as it appears to us, is one of the most certain sources of hospital disease." It is difficult to understand, then, why the cases from which these diseases spring should be all placed together, for such a plan must surely have the result of accumulating an evil influence which must re-act upon every inhabitant of the ward. No one can have visited a large hospital without becoming acquainted with the ever-present hospital smell; and it does not require any unusual keenness of scent to discover that this is always much stronger in the surgical wards than in the medical.

In the report already quoted, we find that "during a severe attack of phagedæna which occurred at the Birmingham General Hospital it was found advisable to do away temporarily with the distinction between medical and surgical cases, in order to separate from each other, as far as possible, the cases of open wounds." Would it not be better to employ this method of arrangement as a constant preventive rather than to keep it in reserve merely as a remedy in desperate cases? and can the staff of the Birmingham General Hospital show any good reason for having gone back to the old-fashioned, and confessedly dangerous, custom of congregating the surgical cases? I can see no reason for the collection of a number of cases of open wounds in the same room beyond a very slight addition to the convenience of the medical staff: a point which I am sure every surgeon would at once concede in order to obtain an advantage for his patients.

It is a matter for very careful inquiry how much the mixing of the cases which necessarily obtains in small hospitals may have conduced to their lower mortality. If this point be found to be of as much importance as I suspect it

is, it will follow that the use of special wards for accidents
and operations is a mistake; and it will have also to be
considered—as indeed it has already been by very eminent
authorities—whether, with proper precautions, certain zymotic
cases may not be more safely treated in the general wards
than in special rooms set apart for them, in those cases where
it is impossible to place them altogether apart; and this
question need, of course, only be discussed where the number
of such cases is comparatively small.

It is a very remarkable fact that though the evil effects
of crowding human beings together was recognised many
centuries ago, and though every now and then some sufficiently shocking illustration of them has occurred, we are
not even yet fully impressed with its dangers. Between
such extremes of experience as the Blackhole of Calcutta,
and the morning headache and loss of appetite which follow
the attendance at a crowded meeting, there is a wide field of
risk, most of the details of which are habitually neglected
save those which stand out in tangible relief. Year after
year our population is decimated by diseases which are the
result solely of our living too closely together; and it is, I
fear, beyond a doubt that we send hundreds of patients into
our hospitals every year to die who would come out cured
if we managed these institutions better.

Dr. Bristowe and Mr. Holmes tell us that " the general
aims and methods of treatment do not appear to vary in
various parts of the kingdom, as far as hospital practice is a
test." They further say that, " if our hospitals present one
defect more conspicuously than another to the eyes of an
attentive observer, it is that of overcrowding." These are
conclusions to which I think no exception can be taken.

There can be no doubt that special skill in the diagnosis and treatment of particular diseases, together with certain most important advances in surgical practice, have had a material and even tangible effect in prolonging individual lives ; but very few curative efforts appear as yet to have had, and perhaps never may have, any perceptible influence in advancing human longevity. In order to determine the existence of any such influence it would be necessary first of all to have a more accurate knowledge than we yet have of the natural history of disease.

But to determine the effect of any hygienic condition is a much more simple matter ; and one of the most important services to which a Government could direct the machinery of a scientific commission would be an inquiry into the influences of varying amounts of floor area and cubic space in hospitals. We have only to look at the results of the Leeds Infirmary, the Liverpool Southern, and St. Bartholomew's Hospital, to be assured that the words quoted above in reference to overcrowding are not at all too strong. In the Supplementary Report for 1875, Dr. Farr makes it clear that there is a definite relation between the density of the population of a district and their death rate, the latter increasing in the ratio of the sixth root of the former. From this formula he found that the estimated mortality differed from the actual death rate only by ·0001 per cent. Knowing this to be the case in a population the great majority of which are in perfect health, we can safely conclude that the rate of increment must be far more rapid in a population of the sick and hurt. It follows also, not only as a probability but as a certainty, that in some special diseases overcrowding must have an especially terrible effect.

The aggregation even of men selected for their perfect

health seems greatly productive of disease ; for Dr. Farr gives in the following table a comparison of the death rate of the army at home with that of the male population at army ages, which displays a dreadful waste of life—an important factor of which must be their aggregation in barracks :—

		Mortality per 1000.
Total Male Population of Army Ages { Country ...		7·7
{ Town ...		9·2
Ditto in Manchester—Unhealthy Area ...	...	12·4
Army { Line Regiments	...	18·7
{ Guards	...	20·4

It should be our first duty, therefore, to ascertain exactly what amount of floor area and cubic space is allowed for each bed occupied in every hospital in the country, and then to see how the distribution of these affects the general death rate. From the experience in the Leeds Infirmary, the only hospital of which I have full measurements, I should be inclined to say that under no circumstances whatever should there be more than one bed for every 150 square feet of ward floor, and that every bed should have a minimum cubic space of 3000 feet. In large hospitals, especially those containing zymotics or surgical cases only, or those built of more than two stories, these allowances should be greatly increased. Inasmuch as we find that piling dwellings on the top of one another, as by flats in Edinburgh and Paris, is a sure way of giving birth to and spreading zymotics, we ought, by legislative interference, to prevent the occupation of any building as a hospital which has more than three stories.

In the *conclusions* of the report by Dr. Bristowe and Mr.

Holmes is to be found a valuable summary of the impressions they gained by the inspection of a large number of hospitals. These conclusions are probably correct in every particular, but it is to be greatly regretted that they did not support them by more complete statistical details. " We may add," they write, " the healthiness of hospitals is less dependent on the form, size, and distribution of wards, than it is on ventilation, drainage, cleanliness, and proportion of inmates to space. A hospital of defective construction may, by careful attention to these latter conditions, be rendered, even in a large town, comparatively healthy ; and a hospital built on the most approved plan, and occupying the choicest site, may be rendered in the highest degree unhealthy by their neglect." Against such opinions no objections can be urged, and out of the long list of hospitals from which I have obtained returns it is unfortunately only too easy to select instances, both in large hospitals and in small, where there is ground to believe that the high death rate is owing to sanitary imperfections. For instance, let us take the example of the hospital at Taunton, where, unless the results could be explained by the existence of some cause beyond the control of the management, they were so bad as almost to justify a judicial investigation. In Bristowe and Holmes' report we are told that in 1861 an outbreak of erysipelas occurred in this hospital which lasted for six months, and during that time 46 cases were attacked, all the surgical wards being invaded by the disease. Nearly all the operation cases were affected, and two died. My returns for the last six years show that the mortality is 5·32 per cent., whilst the mean residence is 53·8 days, from which I conclude that the hospital is not efficiently managed, and that the same causes still exist, though perhaps to a less extent, to which the outbreak of erysipelas might.have been attributed. We may also fairly suppose that the mortality given is much too high for such a hospital, and that it will be perceptibly diminished by the same steps which will curtail the residence. If we compare the Taunton returns with those of Cheltenham, which is next on the list, we find a difference of 15·2 days in the mean residence, and 2 per cent. in the mortality. These two towns have the same general death rate, and very nearly the same average numbers of patients are treated in the two hospitals. We may reasonably infer, therefore, that the difference in the mortality is produced by conditions in the Taunton Hospital which are removable.

There is in existence a very general impression that a hospital becomes more unhealthy as it grows older, but this is not based upon any exact facts, so far as I can learn. It seems rather to be one of those misleading general impressions upon which men are so apt to lean until they are forced carefully to consider their position. There are many hospitals, now nearly a century old, which do not give any indications of increased unhealthiness, which is probably to be attributed to a greater watchfulness on the part of the management; whilst the Manchester Infirmary is only some twenty-five years old.

It is of course more likely that an old house of any kind will become unhealthy by want of care than that a new one will; but mere age should never be accepted as an apology for the bad results of any hospital. There are at least two important hospitals which give very conclusive evidence on this point—the Norwich Hospital and Guy's. In the former, age did not seem to cause any marked variation in the mortality for 90 years, whilst sanitary defects raised it at once 2·2 per cent. In Guy's, the mortality has slowly but steadily diminished—a change which has probably been effected by hygienic improvements.

Numerous instances may be quoted to show that old hospitals have been found to be unhealthy; but before the statement can be accepted that they were unhealthy merely because they were old, we must know whether their arrangements were such as they ought to have been. On the authority of Dr. Guy (*Statistical Journal* for June, 1867), I find that in the old and dilapidated workhouse occupied in the early days of the existence of King's College Hospital, the mortality was 8 per cent. In the new building occupied at present by the same institution, the mortality rose to 10·9 per cent. between 1857 and 1861; from 1861 to 1870 it was 11·557 per cent., and during the last six years it has reached 12·05 per cent. None of the explanations so frequently given in excuse for a high hospital death rate are applicable here. The staff of the hospital is very much what it always has been—one of the most distinguished in Europe. The surroundings of the hospital have not been altered; and it does not appear in any published document that the cases now treated are in any way more severe than they were ten years ago. It therefore rests with the executive to show that the sanitary condition of the hospital is everything that can be desired.

Still another example may be taken from the history of St. Thomas's Hospital; for we find that the mortality is much higher in the new building than it was in the old, and higher even than it was when the hospital was temporarily accommodated in the old theatre at Newington.

There the patients were placed in three huge wards, and
the surgical mortality rose 3·2 per cent. above what it was
in the old hospital, whilst the medical mortality rose only
1·9.   Dr. Peacock, in the *Statistical Journal* for 1867,
endeavours to show that this rise was due to a selection of
the more severe cases, but this is far from being evident in
the facts.   If it were so, the rise in the mortality ought
to have been chiefly in the medical cases; but as it was in
the surgical, it is more than suggestive of an unhealthy
condition of the arrangements, and in my own mind this
is greatly strengthened by my recollection of the want of
light in the huge low-roofed wards, and the inherent
difficulties there would be in ventilating such places.

But what can be given in explanation of the rise in the
mortality in the new hospital, where it is to be supposed
everything has been done which human ingenuity could
suggest to secure the best results?   Can it be possible
that the authorities have, with well-meant but erroneous
intentions of economy, closed some wards and overcrowded
others, instead of diminishing the total number of patients
within the margin of their available funds?   My amputation statistics suggest an affirmative answer to this important
question.

In the paper already quoted, Dr. Guy expresses his belief
that " within the limits of the same capital city the mortality of hospitals is mainly due to the causes which determine the nature and severity of the cases admitted."   If
this be true, it will not be a difficult matter to demonstrate
its proof by taking the cases of four London Hospitals—
as University College, which may be compared with St.
George's ; and St. Bartholomew's, which may be compared

with King's College. In the case of the latter pair, it must be noticed that the St. Bartholomew's district has a general death rate higher than that of the district in which King's College Hospital stands; so that the probability is that more severe cases are treated at Bartholomew's than at King's. At any rate, the contrary must be proved, and not merely stated. The district mortality of University College Hospital is higher than that of St. George's, but the excess is not sufficient to account for the great difference in the mortality of the two hospitals.

In some of the London hospitals the average general population is found to be very unequally distributed through two parts of the year, these being governed by the existence of the session of the Medical School. Bristowe and Holmes especially mention this as being the case at the London and University College Hospitals (Report, p. 465). This must mean that for seven months of the year, and those months when ventilation is least attended to on account of the cold, the wards are far more crowded than during the summer, when ventilation is more free. In the London Hospital the bed margin for the whole year is about 16 per cent., whilst in University College it is only a little over 10; so that overcrowding for half a year would be very easily accomplished, and yet it would be very difficult to detect in a mere annual statement of the hospital population. It becomes, therefore, a very important question whether any, and if any how much, of the excessive mortality of these two hospitals, and of others as well where the same custom prevails, may depend upon this overcrowding in winter. To fulfil the requirements of a clinical hospital

is a very commendable purpose, but it should be done with
the most careful attention to the interests of the patients
concerned.

There is another matter in connexion with the existence
of a medical school at a hospital which, though its effects
cannot be exhibited by statistics, must present itself to the
mind of every thoughtful hospital surgeon, as probably an
important cause of mortality.   It is almost uniformly the
custom for medical students to be in attendance on hospital
practice during the time that they are engaged in the dis-
secting room ; and with a singular inaptness, the curriculum
is generally so arranged as to place them amongst the very
cases—those in the surgical wards—to which they are most
likely to do harm by carrying infection.   I do not think
that the zymads, which must and do cling to the hands of
the dissecting-room student, could do much harm to a case
of pneumonia ; but that they are and must be dangerous
to a case of amputation, or of ovariotomy, or to a parturient
woman, I think no one will be found bold enough to deny.
It seems to me, therefore, that it is desirable on behalf of
the hospital patients that a stringent rule should be
enforced by central authority, securing that until students
have ceased their study of practical anatomy they should not
be allowed to attend the surgical practice of any hospital.

In order to obtain anything like an exact estimate of the
work done by any particular hospital, there are two elements
in the calculations which must be considered, but for which
we have as yet no sufficient data.

The first of these, and the less important of the two, is

the proportion of the sexes. In a few hospitals, chiefly small ones, where I have been able to take out the admissions and deaths of the sexes separately, I have found that the admissions of men to those of women stand as 59·2 to 40·8 ; and that there were 54·46 deaths of males to 45·54 deaths of females ; that is, that though more males are admitted than females, the ratio of the female deaths is, to their admissions, higher than that of the males. This is unexpected, but it is in close relation with what is given by the Hospital Committee of the Statistical Society to the effect that the male admissions are to the female admissions as 6 is to 4 ; whilst the male deaths are to the female deaths as 5 is to 4.

Perhaps of all the factors in the calculation of hospital mortality the most important is that of age, yet it is the one upon which we have the least information. Bearing in mind the enormous disproportion of mortality which occurs during that period of life which corresponds with the ages admitted into Children's Hospitals, it is at first sight almost a matter of surprise that the highest mortality found in these institutions should be 11·421 per cent., and that the average of the years from 1861 to 1870, during which period the returns are fairly complete and trustworthy, should be so low as 6·07 per cent. The only statement which I have been able to obtain of the relative hospital mortality at various ages is one drawn up by the Hospital Committee of the Statistical Society, and which, though not based upon sufficiently extended observations to make it exact, still may, I think, be taken as representing something pretty near the truth. I have added the last two columns in the following table :—

Hospital mortality at different ages per cent. of the patient
population, with ratio between it and the death rate of
the population of England at similar ages.

Age.	Hospital Mortality.	General Mortality.	Ratio.
0– 5 ...	18·8 ...	6·573 ...	1 to 2·86
5–10 ...	8·6 ...	·737 ...	1 to 11·66
10–15 ...	4·7 ...	·582 ...	1 to 8·07
15–20 ...	2·7 ...	·766 ...	1 to 3·52
20–30 ...	4·6 ...	·948 ...	1 to 4·85
30–40 ...	7·4 ...	1·217 ...	1 to 6·08
40–50 ...	10·1 ...	1·638 ...	1 to 6·166
50–60 ...	14·1 ...	2·703 ...	1 to 5·216
60–70 ...	27·9 ...	5·484 ...	1 to 5·087
Total average 10·88		2·294	1 to 4·742

The Committee also made an attempt to calculate a percentage of mortality for certain diseases at different ages.
This would be a most valuable addition to our knowledge
if it could be obtained, but at present it is beyond our
reach.

The period of life when hospital mortality is lowest is
the quinquenniad from 15 to 20, but this does not coincide
with the lowest period of the general mortality, which is
between 10 and 15. This is probably to be explained by
there being a smaller proportion of the hospital population
in the first than there is in the second of these two age-
periods. Such a difference would also probably explain the
difference in the hospital and general mortality in the first
quinquenniad of life, as babies and very young children are
not usually left in hospitals as patients, so that in that age-

period it is likely that both the actual numbers are less than in the next, and that the diseases which are most mortal amongst young children are not taken to hospitals, both of these factors swelling the external and diminishing the internal mortality. It at once becomes evident how important it would be to obtain for every hospital a mean patient-age, and a mean death-age.

As every human being has to die, the true indication of sanitary advance, of the relative salubrity of various districts, or the success of the practice of medicine, is not to be found in a mere statement of mortality percentage, but in a statement of the mean age of the population and the average age at death. To see this completely it is only necessary to refer to the admirable chapter on the "Effect of the Extinction of any single Disease on the Duration of Life" in the Supplement to the last (35th) Annual Report of the Registrar General. Thus Dr. Farr calculates that if none of our male population died of zymotic diseases, their mean life-term, after birth, would rise from 39·68 to 46·77 years. If phthisis were suppressed, the mean life-term after 35 years of age would be raised by 30·77 years. If we did not suffer from cancer our chances of life at 55 would be 16·25 years longer than at present.

There are other means of testing the results of hospital practice, as by taking groups of cases which are strictly comparable, such as ovariotomies and amputations. The latter has been selected by several writers on this subject, but specially by Simpson. I have been able to collect a large mass of statistical information of this kind, which I have tabulated in the Appendix.

I have collected nearly 7000 amputations, but of these
only 4948 can be used for statistical purposes, being
those from hospitals in whose reports or returns they
are properly classified.   By far the larger number of
hospitals seem to keep no proper record of the work
done in them, and even where it is kept, it is seldom
published in the reports in such a form as that it may be
used for statistical purposes.   The favourite form of table
is one which details the serious operations of the year, but
gives no information as to their results.   In some reports
the results are given, but the amputations are not classified
under the heads of " accident" and " disease."   In others
they are classified under headings which do not bear a
uniform meaning.   Thus, though the word " primary " is
always understood to mean an amputation performed on
account of injury at a period not removed more than a few
hours from the occurrence of the accident, yet the corre-
lative term "secondary" is obviously used in widely different
ways.   Sometimes it refers to an amputation performed a
few days after an accident, sometimes it includes only
amputations for disease ; and still further confusion is introduced by the employment of a third term, " intermediary."
If all amputations were tabulated under the headings
" accident" and " disease," their statistical examination
would be greatly facilitated ; and I need hardly remind my
professional brethren that a careful examination of a bulk
of figures is still necessary to decide some important points
in practice, even in connection with amputation—the most
primitive surgical operation we have.

As far as I can see—but the figures which I can bring to
bear upon this question are certainly insufficient—no very
material difference exists in the results of operations

performed immediately after, and those performed a few days after, the receipt of the injury; so that for our immediate purpose these cases may be classed together. But for the purposes of establishing a rule in practice, it becomes apparent that if it were found to be really the case that these results coincided, secondary amputation for injury could not be considered so favourable as immediate amputation; for it involves the elimination of the worst cases—those for which immediate amputation was deemed necessary. Amputations performed a few days after the injury would therefore form a very fair basis for the comparison of the results of practice in different hospitals, if the returns could be made accurate, and if the numbers were large enough. Neither of these conditions, however, can be at present fulfilled.

I issued a special circular in 1875 for amputation statistics, but I cannot say that my results are commensurate with the amount of labour involved. Neither do I feel at all certain that the accuracy of a certain number of the returns is such as to justify any absolute conclusions. The difficulty of obtaining information of the simplest kind from a number of people, can only be appreciated by those who have tried it. I asked for amputations of limbs, and those only, excluding amputations through the wrist and ankle joint. Yet I got a large number of returns containing almost everything but amputations. From an important hospital a return was sent in which my columns were filled by a most carefully detailed account of thirty-one operations, not one of which was an amputation, but which included cataract, cancer of the lip, and prolapse of the uterus.

A very large number of my returns were, for these and other reasons, useless. In the Appendix I place an analysis of all the information I obtained from returns and from reports, upon which I could place reliance. Although I think I may fairly say that it forms the most important contribution to amputation statistics which I have yet seen, I do not feel entitled to imagine that it displays the absolute results. And I must say further, that it seems to me a matter of the deepest regret—I would almost urge that it is discreditable—that the absolute value of an operation so ancient and so frequently performed as limb amputation should be still incapable of demonstration.

In dealing with amputations, it becomes at once evident—and a glance at my summary (Table A) will establish this—that they must be divided into two great classes, according to whether the reason of the operation is an accident or a disease; because the total mortality in the former case stands to the same in the latter as 32·8 is to 22·22.

All amputations for accident are fatal once in 3·05 times, whilst all amputations for disease are fatal once in 4·5 times; and it certainly is remarkable that the numbers dealt with in the two cases are nearly equal, the accident amputations having only a fractional excess ; and the largeness of the numbers employed are sufficient, I think, to remove this from the chapter of mere coincidences.

Besides this initial division, it becomes evident that the operations must be classified according to the limb affected. Amputations through the femur are found to be by far the most serious ; and if the limb is wholly removed at the hip joint, recovery is so exceptional that I have uniformly eliminated this amputation. The same remark applies to double primary amputations, for the recoveries are so few, and the deaths so numerous, that to include them in the returns would be greatly to injure the argument. For the

TABLE A.—Summary of 4948 Amputations.

Hospitals	ACCIDENT								DISEASE								TOTAL	
	Thigh R.	Thigh D.	Leg R.	Leg D.	Arm R.	Arm D.	Forearm R.	Forearm D.	Thigh R.	Thigh D.	Leg R.	Leg D.	Arm R.	Arm D.	Forearm R.	Forearm D.	R.	D.
Hospitals under 20 beds	15	12 / 44·44	52	16 / 23·52	34	7 / 17·09	17	1 / 5·88	26	8 / 23·29	18	1 / 5·5	18	2 / 11·11	8	0 / 0	188	47 / 20·
20 to 99 beds	44	28 / 38·9	76	30 / 28·3	65	10 / 13·33	54	1 / 1·88	65	24 / 27·	53	9 / 14·53	30	1 / 3·22	15	1 / 6·6	402	104 / 20·57
100 to 199 beds	54	44 / 44·9	94	51 / 35·17	103	26 / 20·16	76	5 / 6·17	153	51 / 25·	124	16 / 11·42	53	6 / 10·1	35	5 / 12·5	692	204 / 22·77
200 beds and over	238	223 / 48·37	282	207 / 42·37	311	134 / 30·12	198	39 / 16·47	687	274 / 28·56	337	87 / 20·53	122	31 / 20·2	124	17 / 12·04	2299	1012 / 30·58
Total	351	307 / 46·66	504	304 / 37·62	513	177 / 25·64	345	46 / 11·76	931	357 / 27·7	532	113 / 17·54	223	40 / 15·2	182	23 / 11·23	3581	1367 / 27·68

Total Mortality after Amputation for Accident, 32·8 per cent.
    „   „   „   Disease, 22·22 „

TABLE B.—General Statement of Amputations in the Four Groups of Hospitals.

	ACCIDENT				DISEASE				TOTAL		
	R.	D.	Total.	Mortality per cent.	R.	D.	Total.	Mortality per cent.	R.	D.	Mortality per cent.
Hosps. under 20 beds	118	36	154	23·4	70	11	81	13·58	188	47	20·0
20—99 beds	239	69	308	22·42	163	35	198	17·7	402	104	20·57
100—199 „	327	126	453	27·7	365	78	443	17·6	692	204	22·77
200 and over	1029	603	1632	37·	1270	409	1679	24·4	2299	1012	30·58
Total	2547		2547	32·8			2401	22·22	3581	1367	27·68

TABLE C.—*Relative Amputation Work of Eight Large Hospitals.*

	ACCIDENT.		DISEASE.		TOTAL.	
	Annual Number.	Mortality per cent.	Annual Number.	Mortality per cent.	Annual Number.	Mortality per cent.
Glasgow Roy. Inf.	40	40	24·4	16·31	64·4	30·77
Leeds	34·5	20·28	23·5	17·	58	19·
Guy's	26	46·4	31·	28·67	57	36·76
St. Bartholomew's	13	30·3	23	25·	36	26·7
London Hospital	13·3	52·6	8	56·5	31·3	54·7
Birmingham Gen.	16	35·02	15	20·16	31	34·49
St. George's	6	54·17	15·5	34·7	21·5	40·
St. Thomas's	9·6	44·8	7·3	28·8	17	37·87

TABLE D.—*General Comparison of Three Large Hospitals, and the Mean of Three Small District Hospitals.*

	Mean Residence.	Mort. per cent. of all Patients.	AMPUTATIONS.												TOTAL.	
			ACCIDENT.						DISEASE.							
			Annual Number.	Mortality per cent.					Annual Number.	Mortality per cent.					Total.	Mort. of all Amputations.
				Thigh.	Leg.	Arm.	Fore-arm.	Total.		Thigh.	Leg.	Arm.	Fore-arm.	Total.		
Leeds	25·62	6·78	34·5	32·36	20·75	12·5	7·9	20·28	23·5	23·6	11·76	11·11		17	17	19·
Birmingham Gen.	27·8	8·07	16·	50·	42·33	25·64	20·49	35·02	15	21·9	19·6	22·22	16·6	20·16	20·16	34·49
St. Thomas's	36·35	12·13	9·6	59·17	47·05	44·44	15·38	44·8	7·3	38·46	16·6			28·8	28·8	37·87
Mean of Walsall, Dudley, and West Bromwich District Hosps.	27·6	7·39	12	26·66	26·3	11·77		19·3	5·3	20	11·11			12·	12·	17·

converse reason, amputations through the wrist and ankle joint are not taken into account, though Mr. Callender seems to think they should be.

Glancing at the general results, it certainly is somewhat disappointing to find that of *all* the amputations performed in hospitals, more than *one in every four dies;* and that even of *all* the amputations for disease, the results are not so good as to secure the recovery of four out of five! Knowing what has been done for ovariotomy, which surely must be regarded as quite as serious an operation as any form or kind of amputation, it is not, I think, too much to believe that the amputation mortality displayed in my tables might be and ought to be greatly reduced.

Of all the amputations, that from which the least definite conclusions can be drawn by a comparison of the death rate, is certainly primary amputation of the thigh. Here the shock and mutilation are so great, and the chances of other injuries so constantly present, that the figures possess no remarkable value. Still it must be noticed that the death rate in this amputation is higher in the hospitals having more than 200 beds than it is in any of the other classes; and there can be but little doubt that if a patient should recover from the immediate effects of this operation, he is more open to hospital influences of a septic character than if he had had any of the other amputations performed.

In primary amputation of the leg the conditions for contrast are far more decisive, and the numbers employed in each of the four hospital groups are sufficiently large to be statistically valuable. The operation is an extremely serious one, and in the large hospitals it is almost as fatal as amputation of the thigh, but in the small hospitals it has very little more than half of that mortality. Besides, there

is a continuously rising death rate in the four groups, which
is as suggestive as anything well can be, that Simpson
was right when he said that amputations were fatal in a
direct ratio to the size of the hospital in which they are
performed.   I have already said that I do not believe that
this is a question of mere size; but I do believe it is due to
causes inherent to increase in size, and which would be best
removed, and might in the future be wholly prevented, by
diminishing the number of patients treated in the large
hospitals.

In primary amputation of the arm the mortality in the
large hospitals is nearly double that of the other three
groups when united, but it is perhaps in the removal of the
forearm for accident that the result of hospital influence is
most visible.   In hospitals having less than 200 beds, 147
of these amputations were performed with only 7 deaths, or
less than 5 per cent.; whilst in the large hospitals, of 198
operations 39 died, or 16·47 per cent.: an enormous and
wholly inexcusable increment.   If we are told that this
may be explained by a more serious character in the cases,
or a worse condition of the patients treated in the large
hospitals, we may fairly demand some very substantial proof
of a statement intrinsically so improbable.

Amputation of the thigh for disease is an operation
which has a mortality singularly free from fluctuation; but
even here the large hospitals have the worst of it.   Whether
the mortality from this operation might not be greatly reduced in hospitals of all sizes, is a question which hospital
managers might well ask themselves.   At any rate, in this
operation the evil effects of hospital influences are not
decidedly manifested.   In amputations of the leg, arm, and

forearm, however, it becomes apparent that increase in size of the hospital means a diminution of the chances of the patients' recovery. And I must once more urge that it lies with the hospital authorities to show that this is inevitable.

In Mr. Callender's paper on Amputation Statistics, in the fifth volume of St. Bartholomew's Hospital Reports, an attempt is made to show that hospital amputations are not so bad as was represented by Simpson. The figures used, however, are not satisfactory; for, in the first place, the numbers used in the construction of the most important tables of the *sorites* are not large enough to be statistically valuable; and in the second place, Mr. Callender uses his figures with a fallacy something like an undistributed middle term. In fact, he does not use them in the same way all through his argument, so that, even if it were sound in other respects, which it is not, it would be unsound in this. I have used my figures so far in a constant form; but even if I diverged, and used Mr. Callender's method of lumping all my amputations together, without reference to the limbs affected, the numbers of each particular amputation, or as to whether the amputations were for accident or disease, I should be able to display a marked advantage for small hospitals over large ones, and a death · rate constantly increasing in proportion to the size of the hospital. The summary in Table B shows this clearly, and the margin of increase in the mortality is far too great to be entirely without value.

But still another inquiry must be made into the comparative amounts of amputation work done in various hospitals, because we have been constantly told that one of the chief reasons for high hospital mortality is to be found in the accidents treated in certain institutions. If this were true we should certainly find the highest amputation mortality coincident with the largest annual number of amputations performed. We also find it constantly hinted that the primary surgery in the London hospitals is larger in amount and more serious in character than in the provincial hospitals. To display the inaccuracy of such ideas I have constructed a table from the amputation statistics of eight hospitals from the fourth group. These are selected because they are the only institutions whose returns are available for comparison (Table C).

Now it becomes apparent that there are at least two provincial hospitals in which more, and probably more serious, primary surgery is met with than there is in any London hospital. Indeed, with the exception of Guy's, no London hospital seems to have many primary amputations performed within its walls; and in that particular hospital, which is now supposed to be the most important and the best, because it is the newest, St. Thomas's, there are 9·6 primary amputations of all kinds performed every year, as against 34·5 performed in the Leeds Infirmary.

The construction of this Table led me to another (D), which is the last with which I shall trouble my readers on this subject, but I think it is really the most important.

I have selected the three large hospitals for a detailed comparison, because they are all representative institutions, and because of all those from which I procured complete information, they seemed best suited for contrast. The Leeds Infirmary is a new hospital, constructed and managed upon what I believe to be, with some few exceptions, into

which I need not now enter, the best possible principles.
It is the chief hospital of a large consulting area, including
a large manufacturing population. As I have already said,
this new building seems to have exercised a most beneficial
influence upon the hospital mortality of Leeds. That the
amputation mortality has been similarly affected, I do not
know, but who can doubt it?

The General Hospital of Birmingham is, on the contrary, an old building, in the main. Most of its wards
have low roofs, and have always seemed to me overcrowded, and its intrinsic sanitary conditions are very
inferior to those of the Leeds Infirmary. It was
with surprise, therefore, I must confess, that I found that
its general death rate was only 1·29 per cent. higher than
that at Leeds, and that its patients resided within its walls
only 2·18 days longer than did the patients at Leeds in
their splendid new building. But in the amputation death
rates of the two hospitals the advantages of the better sanitary arrangements at Leeds becomes at once apparent.
The number of primary amputations at Leeds is more than
double that at the Birmingham General Hospital, yet their
mortality is little more than half; and in the instances of the
leg, arm, and forearm, it is considerably less than half. The
only case in which Leeds has the higher mortality is in
amputations of the thigh for disease, and the difference
there is only fractional. The floor space allowed in the
two hospitals for each patient is about equal, but the cubic
space is nearly 45 per cent. higher at Leeds than it is at
Birmingham. The difference in the amputation mortality
of these two hospitals is very striking, and is very suggestive
of the high value which, in some instances at least, must

be placed on differences in the general mortality, which at first sight seem trifling.*

By the courtesy of the Managing Board of the General Hospital at Birmingham, I was enabled to take out their statistics myself, so that I know they can be subject to only very slight correction, chiefly due to a very small number of amputations whose results were not recorded. In looking over the operation books, it became quite evident that during the last four or five years the number of primary amputations—indeed, of operations generally— has very much diminished, and this fact finds a very ready explanation in the establishment of a number of small district hospitals—chiefly those of Walsall, Dudley, and West Bromwich. The last line of Table D is formed by a mean calculated from the returns of these three hospitals, save that the item of residence is estimated and not actual, the Dudley Hospital not having made a return for residence. It will at once be seen, and I hold it to be proved incontestibly, that a marked saving of life has been effected by the transfer of the amputation cases from the large to the small hospitals. In the case of primary amputations, part of this may be due to the fact that the cases are not removed to a distance to be treated, though it must be remembered that this argument cuts both ways, for the

---

* The superiority of the results obtained at the Leeds Infirmary over those at the other two large hospitals is again shown by comparing their ovariotomies. The numbers are not large enough to give them absolute value, but they strongly support my general conclusions :—

	Cases.		Deaths.		Mortality.
Leeds Infirmary	59	...	30	...	50·86
Birmingham General	13	...	9	...	69·44
St. Thomas's	27	...	17	...	62·88

The effects of hospitalism in this operation are sufficiently startling.

worst cases are eliminated by the removal, either dying on the journey or being so bad when admitted that no operation is attempted. A marked improvement is, moreover, evident in the general death rate and in the amputations for disease, so that it is highly probable that an immense gain would accrue if the General Hospital of Birmingham were either rebuilt, as the Leeds Infirmary has been, on a better plan and on a better site, or, better still, if it were to be broken up into three or four smaller hospitals.

As to the results displayed by the returns from St. Thomas's Hospital, unless they can be satisfactorily explained as being due to some inevitable and irremovable conditions, I must say that I have grave doubts about the advantages gained by the populace of London from the palatial edifice on the Albert Embankment.

The whole of my statistics tend to prove that after the number of beds in a hospital exceeds 100, the risks to life become so much increased that it is questionable whether any hospital should be of larger size than this. If circumstances make it necessary that the hospital should be larger, most undoubtedly special arrangements and precautions should be taken to obviate the extra risk which is involved.

But whatever objections may be urged against the deductions I have advanced, based on an examination of amputation mortality, they cannot be held for a moment when we consider what can be said of another operation of much more modern date.

It is no part of my business here to enter into the history of ovariotomy, though none of the records of surgery are more interesting ; but one phase of its history is of especial importance for this inquiry. When we look

back on the long and bitter discussion which preceded the
establishment of ovariotomy as a legitimate surgical
operation, and when we examine the two parties in it, we
must be struck by the fact that the opponents who spoke
most weightily against it were surgeons who were attached
to large hospitals, who had tried the operation there, and
who had failed in securing any reasonable amount of
success. On the other hand, the men who argued for the
operation, who had tried it and succeeded, were practitioners
like Henry Walne and others, who performed the operation in private houses or in small hospitals. It may
be said that the establishment of ovariotomy is mainly
due to the success of two men, and both of these men did
their work in small hospitals; the late Mr. Baker Brown,
in the small hospital known as the London Surgical Home,
which he established for the purpose, and Mr. Spencer
Wells, in the Samaritan Hospital.

In an operation like this, there can be no doubt that
special experience in its details must greatly contribute to
its success. It is an operation far more full of risks than
any other surgical proceeding, and one where inattention
to the smallest details may therefore have the most disastrous results. The enormous experience, amounting to
more than eight hundred cases, which Mr. Spencer Wells
has had, must now contribute largely to his success. But
there is an element of much greater—indeed, of overwhelm-
ing—importance, which directs the results of this operation,
and which points to a conclusion for all other operations
in a perfectly irresistible way.

It must be borne in mind that in the great majority of
the women submitted to ovariotomy, the operation must be
regarded as analogous to a primary amputation. They are

going about, eating, sleeping, and possessing ordinarily good health. They are placed on the operating table, and then submitted to an operation which, even under an anæsthetic, often produces such shock as to send their general temperature down as much as four degrees.

But whether they are in such a condition as that the operation may or may not be regarded as primary, before it is done they are all—with a few exceptions to which no allusion may be made here—in so much the same condition, that one hundred cases will present almost exactly the same risks as another hundred, provided the surrounding conditions of the two sets are alike. If the operators are equally careful and attentive, no differences in the intrinsic conditions of the operation will much affect the result ; for I have found in my own experience, now somewhat large, what all other experienced operators have also found—that difficult, complicated, and apparently hopeless operations often do well, whilst the simple and easy very often go speedily wrong.

Mr. Spencer Wells performed his first operation in February, 1858, and up till October, 1860, he had performed twelve operations in the small Samaritan Hospital, of which eight recovered. During the same time it is known that at least ten operations were performed in the metropolitan hospitals (*British Medical Journal,* December 1868), with only one recovery, and that, remarkably enough, was performed in the small Metropolitan Free Hospital. During that time Mr. Wells' special experience was no greater than that of the other operators, and it is quite enough to read the list of their names to be convinced that in every case a full amount of care and surgical skill was given to the operation. To what then may the difference in the results be attributed? The answer is given in the general results obtained from that time to this, and may be summed up in the words—segregation of the patients.

Before entering on this inquiry, I must again reiterate my regrets concerning the state of our hospital statistics. We can only be certain of one thing about the results of ovariotomy in general hospitals—and that is, that we know all the successful cases. But terrible though the list of fatalities is, it is not at all certain that we have a complete account of them.

In a paper by Dr. Skölberg, of Stockholm ("Om Ovariotomi," 1866), the following table of the results of ovariotomy in large London hospitals is given, together with the authorities upon which the statements depend:—

Hospital.	Cases.	Recoveries.	Deaths.	Mortality per cent.
Guy's	54	33	21	38·8
Middlesex	8	1	7	87·50
King's College	7	1	6	85·71
University	5	1	4	80·
St. George's	7	2	5	71·43
St. Bartholomew's	12	4	8	66·67
Total	93	42	51	54·94

The low mortality at Guy's, Dr. Skölberg explains by the greater precautions taken there, but still the contrast is very unfavourable:—

	Cases.	Recoveries.	Deaths.	Mortality per cent.
Five large Hospitals	39	9	30	76·92
Guy's	54	33	21	47·73
Samaritan (up to Feb. 1863)	106	76	30	28·30

I have collected 271 cases of ovariotomy performed in hospitals having more than 100 beds. Of these, 58·1 per cent. have died: a mortality worse than that displayed by Dr. Skölberg's statistics. Mr. Spencer Wells and Dr. Keith have already proved as fully as any fact in statistics can be be displayed, that in a small hospital the mortality from this operation should not much exceed 28 per cent., and that in private practice it probably would be less than 20 per cent. These figures have convinced me that this operation should not be performed in a hospital, in the ordinary sense of the term, of any kind whatever; and I think that the most enthusiastic conservative will hardly dare venture to support its performance in large hospitals.

This wonderful difference in favour of the Samaritan is not to be explained entirely by Mr. Wells' special experience, for that does not seem to bring its influence to bear fully on his statistics till between the second and third hundred cases. He gives them as follows :—

		Recoveries.	Deaths.
First hundred	. . .	66	34
Second ,,	. . .	72	28
Third ,,	. . .	77	23
Fourth ,,	. .	78	22

It is of course very likely that a large share of this increasing success is due to an increase in the stringency of the precautions which Mr. Wells took in isolating his patients; but some of it must be due to the wonderful dexterity of manipulation which he has attained.

Taking, however, the same experience as applied to cases in private practice—where, of course, isolation could be made complete—and to cases in the Samaritan Hospital, where, though it could be carried out very well, but not completely, we find that Mr. Wells has had the following experience (" Diseases of the Ovaries," 1872) :—

	Recoveries.	Died.	Mortality per cent.
Total Hospital cases, 240	. 176	... 64	... 26·66
Total Private cases, 260	. 197	... 63	... 24·23

These two sets of cases are in every way fit for comparison, and the numbers are sufficiently large and equal to eliminate error, and the conclusion is inevitable that even with the very perfect isolation carried out under Mr. Wells' personal supervision at the Samaritan Hospital, where no infectious cases are admitted at all, and where the number of patients gathered together is small, there is an advantage on the side of complete isolation. In the Samaritan, each ovariotomy case is kept completely isolated in a room by herself during the critical days after the operation, yet the mere proximity of other patients seems to send up the risk nearly two and a half per cent.

This is still better seen when we compare the statistics of another eminent operator, Dr. Keith, of Edinburgh, who rigidly isolates every patient he operates upon, and who has done only one operation in a general hospital, that one having had a fatal result.

Of his first hundred cases only nineteen died, and of his last fifty he has lost only six. Of the first series he says (" Ovariotomy," Edinburgh, 1870)—" Of the whole number of operations, 70 were treated in the same room. Of these, 60 recovered. Nearly all the worst operations were performed there. The greater number of those who died were poor, worn-out women, who came late in the disease.

The mortality would probably have been much lower if there had been earlier operation in many of them."

That fifty women should have been subjected to such a serious operation as ovariotomy with only six deaths, is a fact unparalleled in the annals of surgery; and contrasted with the same operation performed in large hospitals, with results almost converse, it leads us but to the one conclusion—that to perform ovariotomy in a large hospital is an utterly unjustifiable proceeding ; and I am almost even prepared to denounce its performance in a hospital where any other kinds of cases are admitted. My own experience shows a difference of 10 per cent. in the mortality of cases performed in private, and those in a hospital with only seven beds.

From this special operation to operations in general it is not a difficult task to argue. There can be no doubt that in ovariotomy there are two conditions which render the effects very evident—the primary nature of the operation, and the opening of the large lymph sac of the peritoneum. But the statistics of primary amputations, already considered, show markedly that they also suffer from association with patients, and that they are benefited by isolation.

There can be no doubt, also, that conditions which influence operations where the peritoneum is opened, cannot be entirely absent in any operation where an abrasion of tissue is made. This is proved beyond a doubt by the disastrous records of such a state of matters as Miss Nightingale disclosed at Scutari Hospital in the early part of the Crimean campaign, when wounds of all kinds took on a gangrenous action. This gangrene used to be common in our civil hospitals, and is too frequent even now, and

K

goes by the name of "hospital gangrene." It may affect wounds of the most trifling nature. No reasonable person now doubts that it was and is due to bad sanitary arrangements; and it is a very legitimate conclusion that the same influence will do harm to an extent perceptible only in general results and not in local indications, where its originating causes have been modified but not removed. That is, it is certain that a badly constructed or badly managed hospital will give bad results, even when it is not sufficiently unhealthy to be constantly exciting "hospital gangrene" and "hospital fever;" and from the facts of ovariotomy it is equally certain that the nearer a hospital approaches the conditions of an isolated private dwelling in its construction and in the relations of its inhabitants, the better will its results be.

In conclusion, I can only reiterate the opinions of Miss Florence Nightingale and Mr. Cadge, that it would be infinitely better to leave the sick and hurt in their own homes than to place them in buildings where they are exposed to the risks apparent in the returns of certain hospitals.

The whole question is of such great importance that I trust an exhaustive examination of it will be made by a competent and duly authorised body.

APPENDIX.

Appendix.

AMPUTATIONS.

From No. 1—16 the percentage of Mortality is given below each Death Figure.

No. of Hospital in previous returns	Years of which return is made	ACCIDENT								DISEASE							
		Thigh R.	Thigh D.	Leg R.	Leg D.	Arm R.	Arm D.	Forearm R.	Forearm D.	Thigh R.	Thigh D.	Leg R.	Leg D.	Arm R.	Arm D.	Forearm R.	Forearm D.
1	1863—75	21	18 (46·29)	29	24 (45·45)	33	6 (15·38)	35	3 (7·9)	123	51 (29·4)	60	20 (25)	18	2 (10·)	24	1 (4·)
2	1861—74	52	63 (54·8)	70	70 (50·)	51	26 (33·78)	22	10 (30·3)	163	84 (34·)	104	25 (19·38)	27	12 (30·76)	17	4 (19·04)
3	{ 1865—67 and 1868 }	6	9 (60·24)	3	8 (72·72)	7	4 (36·36)	3		9	10 (52·6)	1		25	2 (7·4)	15	2 (11·77)
4	1867—75	56	53 (48·6)	48	33 (40·8)	80	50 (36·36)	33	8 (19·5)	114	30 (20·83)	36	9 (20·)	7	2 (22·22)	6	
5	1866—74	9	13 (59·17)	18	16 (40·8)	10	8 (38·46)	11	2 (15·38)	24	15 (36·46)	10	3 (7·69)	5	3 (37·59)	14	4 (22·22)
7	1865—72	6	11 (64·72)	6	5 (47·05)	6	9 (44·44)	4	1 (20·)	42	25 (37·3)	20	11 (35·4)	8	1 (11·11)	9	1 (6·6)
8	1870—5	46	22 (32·36)	42	11 (20·75)	42	9 (60·24)	35	3 (7·9)	55	17 (23·6)	45	6 (11·76)	2	5 (11·11)	5	
10	1870—5	3	6 (66·6)	2	1 (33·3)	5	12·5	7	2 (22·22)	15	1 (6·25)	1	10 (19·6)	25	16·6	25	5 (16·6)
14	1861—76	25	25 (50·)	49	33·3	58	28·57	35	9 (20·49)	107	30 (21·9)	41	10 (19·6)	1	100	1	100
16	1873—5	5	1 (16·6)	6	36 (42·33)	5	20 (25·64)	3		3	1 (25·)	5	1	4	2	7	2
17	1870—75	4	2	5	1	6	1	5	1	8	6	12		1		1	
20	1871—75	2		1	2	1		5		9	2	2		4	1	2	1
21	1874—75	3	1	3	3	5	3	9		3							
24	1874—75	4	1	13	3	2	3	4		12	2	4	1				
29	1870—75			6		3				11	5	3	1				
30	1870—75	3		6	4	2				4							

Amputations (*continued*).

No. of Hospital in previous returns.	Years of which return is made.	ACCIDENT.								DISEASE.							
		Thigh.		Leg.		Arm.		Forearm.		Thigh.		Leg.		Arm.		Forearm.	
		R.	D.	R.	D.	R.	D.	R.	D.	R.	D.	R.	D.	R.	D.	R.	D.
112	1871—75	2	1	1		3		1		2		1		2			
113	1870—75	2	2	3	1	3	2	2		4	1	8	2	2		2	
114	1870—75	2		5		5	2	1		2	3	1	3	3		1	
116	1870—75	1		2		2		2		2	1	3	2	2		1	
119	1870—75	1	1	1	1	2		1		5							
121	1873—75	3		3		2		9		4		3		1			
124	1875	2	1	2	1	2		1		1		3		1			
126	1870—75			6	3	1	1			1		2					
127	1870—74		2	1		6	2	2		2	2	2	2	1		1	
128	1873—75	2	1	4		2		1		1	4	1					
132	1870—74	2	2	1		6		2				4					
133	1871—74	6		4	1	3	1	4		3	1	1		1		1	
134	1871—74	4		7	1	10	2	1		2		1	1			1	
137	1870—75			3	2	3		2				2					
138	1870—75			1	1						1			4			
140	1873—75									3							
141	1870—75	2		8		2		4	1	2		2					
142	1870—71											1					
143	1870—75	4	2	4	1			2		3	1						
144	1872—75			6	1			2		1							
145	1873—74		2	9	2	6		2			2	1					
146	1870—75		1	2	3	7	1	1				1		2			
147	1870—75						1									1	
151	1870—75	3	1	4			1			1		5				1	1
153	1873—75											2					
157	1870—75									2							
158	1870—75									5							
162	1871—75									2				1			

| | 1872—75 | 1872—75 | 1870—75 | 1870—75 | 1870—75 | 1870—75 | 1870—75 | 1870—75 | 1871—75 | 1872—75 | 1873—75 | 1870 | 1871 | 1870—75 | 1870—75 | 1870—75 | 1870—75 | 1871 | 1872—75 | 1875 | 1873 | 1875 | 1870—74 | 1872—75 | 1870—75 | 1872 | 1873—74 | 1874 | 1871—75 | 1870—75 | 1872—75 | 1870 | 1872 | 1873—75 | 1872 | 1871—75 |
| | 163 | 164 | 167 | 168 | 169 | 171 | 171B | 172 | 173 | 174 | 180 | 185 | 187 | 188 | 190 | 191 | 193 | 194 | 196 | 198 | 199 | 200 | 202 | 203 | 204 | 205 | 208 | 215 | 216 | 220 | 221 | 227 | 230 | 233 | 234 | 235 | 236 |

OVARIOTOMIES.

No. of Hospital				Cases.		Deaths.
1				38	...	24
2	.	.	.	93	...	50
4	.	.	.	27	...	17
5	.	.	.	27	...	17
7		.	.	10	...	7
8	.	.		59	...	30
14				13	...	9
21	.			2	...	2
24			.	2	...	1
				271		157

THE END.

LONDON:
SAVILL, EDWARDS AND CO., PRINTERS, CHANDOS STREET,
COVENT GARDEN.

9 783337 161866